Peddler of Death

AF799909

Mars Knight

ISBN: 978-1-959303-45-9

Dedication

To my late, beloved wife Lucretia, who taught me to appreciate the blessings in my life because, '...things could always be worse'. She inspires me daily.

6/26/1956 - 9/20/2014

Contents

Preface

A flaming object falls from the evening sky to crash into the high desert in New Mexico Territory, and peddler Jethro Kayne finds a handful of injured black cricket-like creatures outside the metal can they arrived in. He cares for and feeds the ravenous clicking bug things, which reproduce rapidly and have an insatiable appetite. Jethro learns to communicate by Morse code with the little monsters he dub clickets. The peddler alienates his pregnant daughter April when his pets kill and devour her lover, and she flees to anonymity in the town of Coyote Gulch, newly minted as Sarah Peale.

General Kayne leads his clicket army, which grows by orders of magnitude to fresh meat at a miner's camp, a town of ruffian outlaws, and a cattle ranch, laying waste to every living creature, stripping every scrap of meat from the bones. Coyote Gulch is on the future menu.

An aimless drifter, a renegade Navajo, a mute boy, and a chronic drunk are drawn to Coyote Gulch by various forces to join the townsfolk in defending against the invading clicket hordes. The battle is enjoined, and Kayne, though severely wounded, senses victory, but his daughter hopes to have the final word, and Kayne never reckoned on the whiskey.

Chapter 1

It was near about sundown on a warm June evening in northwestern New Mexico Territory when Jethro Kayne saw and heard the fireball whoosh through the sky, right over his head! It crashed into the high desert sands of the Colorado Plateau, no more than a half mile from the dusty tracks he traveled over. The forty-eight-year-old peddler reined his mule, in and brought his wagonload of household goods to a stop, about four miles from his shack of a home.

Jeth didn't know what he'd seen, probably a meteor, but perhaps there was money in it, and he wasn't going to pass up the possible opportunity to cash in. He'd been on the road for three tiring weeks straight, selling and trading goods in lonely outposts and settlements strung about the county, and was sick and tired of barely making ends meet. His daughter April bought home more money by washing laundry at the copper mining camp that he'd passed two miles back, than he ever did from trading and selling junk. Of course, she didn't waste money on whiskey, gambling, and whores either. Her loss.

He walked over the sand and around the mesquite and cacti to his hoped-for treasure, carrying an empty flour sack. The object was glowing red at the bottom of a shallow crater when Kayne found it. It wasn't a meteor! The metallic object, pockmarked as if dented by hail, resembled a whiskey keg but only half the size. Jethro Kayne wasn't school-learned, but he'd been around and had seen things. However, the peddler could barely speculate on the origins of this strange item.

If it was manmade, how did it fall from the sky? Maybe it was some trick, like from a traveling circus, shot from a cannon? On the other hand, if it came from outer space like a meteor, who made it? Jeth felt certain there was a fortune to be made, but fear and heat held him back from touching the thingamajig.

The peddler bent over to peer more closely at the big, dented, metal bullet when suddenly a hatch popped open on one side! Jeth fell backward in surprise and sat in the sand dumbfounded as he watched black bugs with dark green-yellow heads emerging from the opening. His mouth hung open in astonishment. The creatures appeared injured, some limped, crawling on only three of their four legs, and one of the things was carried out by two others! They held their fellow bug in tentacle-like appendages, laid it down, and then collapsed next to it.

Kayne stood up and counted nine of the black bug things. They resembled large crickets, with long hopping legs in the back but four tentacles rather than four wings along their sides. He noticed that one was missing the middle of its three eyes and wondered how many dead bug things remained inside the ship.

Jethro grew up on a farm in Missouri, and he knew these weren't locusts or any kind of bug. Overlapping scales covered their bodies, like on snakes, and their four stout legs more resembled those of a small toad rather than the twiggy limbs of the six-legged grasshoppers he was familiar with.

Whatever they were, Jeth saw gold in them; well, maybe just silver, but he knew the things were special and rare, whatever they might be. Maybe they'd escaped from a sideshow or traveling circus, and there'd be a reward! One of the nine, with a red asterisk on its head, looked up at Jethro. It audibly clicked its jaws together in a pattern of some sort, as if trying to communicate. Then it crumpled and fell. The nine pitiful creatures lay in the sand, unmoving.

The peddler carefully picked up the thumb-sized critters, cupping them in his palm, one at a time, and tenderly placed each inside his flour sack. He toted the ailing aliens to his wagon, moving gingerly as if he were carrying nitroglycerin. He set the bag of bug things up on the driver's bench, untied his mule, climbed up next to his prize, and set off for home. He vowed to return to this place tomorrow to salvage the can the critters arrived in when it would be cooled down. That had to be worth a lot, too, to somebody. In eighteen years of peddling goods, Jethro had learned much; there was always another sucker waiting to be swindled. It was all about the spiel.

Twenty minutes later, Jethro was home, inside his two-room shack. The peddler set the nine cricket-like creatures down, onto an old worn shirt, at the bottom of a wooden shipping crate. He set a tin of water next to his new guests and tossed apple slices, some stale bread crusts, and a strip of beef jerky into the box, thinking to offer them a wide choice of foods.

The red-starred individual clicked, while looking in his direction, and Jeth felt certain it wanted to communicate head lay prone but clicked weakly several times, and Kayne dubbed them 'clickets.' He couldn't wait to share the exciting discovery with his daughter. April Kayne laundered for the fifteen or twenty men in the mining camp six miles southeast. She usually spent two weeks at a time cleaning and sewing for them, sometimes longer, and then she'd come home for three or four days to try to recuperate. He didn't know when she was due home next, but he'd learned to determine her day of departing by the thickness of the dust on the chairs and tables. He judged that she'd been gone three days now.

Next morning, Kayne checked in on his patients first thing, and was happily amazed to see all of the food consumed and eight clickets crawling and hopping about in the crate; the ninth was missing! He guessed that the clickets ate their deceased. They'd healed overnight! Jethro looked down at them and said, "Good morning, clickets! Glad to see you're feeling better." He laughed.

A flurry of clicks answered him from eight competing critters, but after a moment, only the red-starred individual clicked its jaws. It reminded Jethro of his time as a telegraph operator in his mid-twenties and the tapping of Morse code. For the moment, he assumed that the clickets hungered again, so he gathered five eggs from his chickens' nest boxes under the lean-to on the east side of his hovel.

After hard boiling the eggs and peeling them, he fed three to his thriving pets and ate two for breakfast. Jeth watched in fascination as the eight clickets tore into the chicken eggs. Their powerful four-way mandibles ripped out great chunks, and Kayne imagined they could bite off half a finger with one chomp! The ravenous little monsters devoured the three eggs in less than ten minutes. They pooped almost continuously, leaving flat black mud specks dotting his shirt on the bottom of the crate like hundreds of tiny ticks.

The red starred individual clicked, and Jeth felt certain it wanted to talk to him while looking in his direction. He considered that whether his clickets came from a secret laboratory or outer space, they must be intelligent. Kayne tested his hypothesis.

He tapped 'Jethro' in Morse code on the side of the wooden crate while pointing at himself with his other hand. The spokes-clicket pointed two tentacles at the gray-whiskered peddler and repeated his name, clicking its jaws together in code! Kayne forgot about the ship in a crater several miles away, and spent hours teaching the clickets English in Morse code.

The creatures learned quickly, usually after only one iteration, and zipped through their lessons on anatomy in a short time; head, arm, leg, scales, tentacles, black, skin, hair, hand, foot, and decimal numbers one through twenty. The Kwarliks used base-eight mathematics, but adeptly assumed the new arithmetic. In no time at all they mastered Morse code. Kayne learned the queen's name was Qweequa and that six of the clickets were female, and two were male; clickets were matriarchal.

Day 6: The insatiable clickets lived outdoors now, and were currently devouring his last chicken. Jethro wondered how he could afford to keep them! Thankfully, there were only eight of the voracious things, but he needed to find a way to make them pay their own keep. He could create an entertaining show and take his clickets on tour to the big cities, like San Francisco! Then again, it might be better to sell them outright. He hoped that April would return home soon; surely, she'd see the wisdom and value in keeping the creatures. Maybe she'd heard whether a reward was offered for the clickets.

The clickets called themselves Kwarliks, and Qweequa told Kayne that thousands of their ships seeded planets all over the galaxy with members of their rapidly reproducing race, which eats its way across space. She related that her kind could remain dormant in space for millenia, until they reached a fertile world. Jeth couldn't grasp the immensity of it all, but he understood one thing for certain; their loyalty depended on him providing their necessities.

His mule would have to be sacrificed to their hunger next, so for the last time, he drove the beast down to the creek, half a mile south, and filled two casks with water there, before making the mule pack the load home.

Day 7: Jethro Kayne's heart skipped a few beats, and he forgot to breathe. Qweequa announced the presence of three hundred and forty-four eggs! Kayne saw the sticky gray blobs of eggs lining the backs and sides of all of them, the tentacles protectively holding them in place. 'Obviously, the males shared egg toting duties.'. There was no possible way in the world that he could feed three hundred and fifty clickets! It was time for a heart-to-heart click with Qweequa, if clickets had hearts.

Qweequa explained the facts of Kwarlik's life to Jeth, and thoughts of jumping off a cliff, flitted through his agitated mind as he read the words he'd printed. He put them down on paper as Qweequa spelled the letters out, and when the queen clicket completed her speech, Jeth would read the words. She'd said, "Kwarlik females outnumber males four to one and lay sixty to one hundred eggs every five to seven days. The eggs hatch in two days, and our precocious offspring are sexually mature by Earth day number three. Kwarliks live hard and fast, and our metabolism is on fire for only about a thousand of your days, by which time hundreds of billions of us will blanket your entire planet.

"My red markings are the sign of a Progenitor, born once every seventy generations or so. My kind lives three times as long as normal because we lay no eggs except under extreme conditions, for I carry the seeds of a new population. Thus, even should every other male and female perish, I can continue the line of descent."

Jethro wondered if he should crush them and their eggs before they turned on him and ate him for supper. Hunger has no morals. He wished April were here! She was only twenty-two but was wise for her years, having grown up without a mother who abandoned the family when April was six. The girl grew up quickly, mostly making her own way, since he wasn't available to her on a reliable basis. Jeth knew that she stayed with him out of a sense of responsibility, and he manipulated her sympathies to his advantage. She owed him.

Kayne's anxiety was quelled after a further conversation with Qweequa; he was excited by her words. She clicked Morse code letters, and he wrote them down, and when completed, it read, 'We owe you our survival and vow to neither eat you nor the daughter you speak of, but as our population increases by orders of magnitude, and as our metabolism requires enormous quantities of nutrients, we are migratory. We are an army that devours all we encounter, and only you know where to find meat. Jethro, we need each other; as an army needs a general, and as a general needs an army. Lead us to food."

Day 11: When he arose, the newly appointed general of the clicket army found more than four thousand clickets devouring his mule. They liked night attacks. The noisy clicking reminded Jeth of cicadas and was audible for many dozens of yards, even though the majority of their numbers were new hatchlings, no larger than a horsefly. Later that afternoon, after he'd finished burying the mule's bones, April arrived home.

She often made the six-mile journey from the mining camp on foot, but this time she rode behind a handsome young fellow mounted on a chestnut mare. Jethro was alerted to their approach by the clickets, which suddenly went silent all at once, and hid away scuttling into the shadows. He initially frowned and watched from the doorway as the man dismounted and helped April down. Anger flushed his face red when they kissed. No one was taking Jethro's girl away; he needed her!

General Kayne asserted his winningest sales technique and quickly concocted a wicked strategy to implement it. "Hallo!" he shouted joyfully as he went out to greet his daughter and meet her friend. April was unused to such emotion but gleefully took advantage and hugged her father. "This is Foster, Pa,' she said by way of introduction. "He chops wood for the miners and hunts and cooks and such."

Pa Kayne shook hands with Foster and said, "Kayne. Hey, why don't you come inside and have a drink for your troubles?" Foster doffed his hat and said, "That's the second best suggestion I've heard all day Kayne, though it weren't no trouble at all." He wound his horse's reins around a worn hitching post and followed Jeth and April into the lopsided shack.

Kayne found a couple of dirty glasses and splashed whiskey into both. The two men had only just swallowed their first drink when chaos erupted outside. A deafening concert of clicks shattered the peace, and a horse screamed in pain and terror!

Foster dashed out the door first, followed closely by April and Kayne. The horse was down but kicking and rolling in the dirt, covered in a black swarming carpet of carnivorous Kwarliks.

Foster ran into the battle, but Kayne stopped on the porch, and he grabbed April's arm before she could enter the fray. Jeth smiled in grim satisfaction as he watched the horror unfold, and his daughter fought and stomped and screamed in a fury, attempting to escape her father's powerful grip.

Foster slapped and swatted at the creatures that swarmed around his horse's head, clogged its nostrils, dug into its ears, and tore its eyes. The clickets turned their attention to their attacker now. Scores of them hopped onto Foster and crawled up his pant legs faster than he could beat them away; for each one he killed, ten more attacked! The chomping monsters penetrated his every orifice and ripped chunks from his face and body. He fell to the ground and rolled in the dirt, trying to dislodge the clinging, biting creatures.

It was unbearable, and April finally escaped Kayne's clutches and ran to her lover, who was literally being eaten alive! She knelt in the dirt and tore at the horrible creatures, clawing them from Foster's body and loudly sobbing. Her father walked up next to her and the writhing man, and mercifully shot him dead.

April shrieked and beat on her father's legs, and screamed at him, "You're horrible! What are those things? What have you done?"

Jethro Kayne firmly pushed April's face away with the palm of a hand and said, "Shut up, little girl and listen to me for once! That's what happens when you whore around like your damn mother did! Look; this is our chance to get out of this shithole and have everything and anything we want! This is my conquering army of clickets, and they go where I lead. I'm tired of being looked down on like I was some kind of insect, so now it's time for the insects to rule, and you and I will be royalty! Don't you see it, April? "

April didn't see it, nor did she want to listen to her dad's story about his army and the promise of unlimited power. She packed her meager belongings and departed her childhood hovel for the last time, leaving her father to try to understand her parting shout. She said, "I don't want to rule the world! I only wanted some happiness, and that can't be found with you."

Jethro figured she was just jealous because doesn't everyone want to rule the world? He would make her pay; he was going to make the whole world pay.

Chapter 2

Red Sky shook his head in consternation; it was just as his series of disturbing visions from a fortnight past had shown him; Death. The copper mining camp was a graveyard. Nothing remained alive. Human, dog, horse, and mule skeletons lay scattered about amongst shotguns and spent shells. Piles of ripped and torn clothes, with bones inside, and saddles girdled around white ribcages, lay outside the deserted tents and wagons of the silent camp. The lonely tents stood tall, waiting for tenants who'd never return; cooking pots sat near cold fire pits, and a few shirts hanging from a clothesline waved in a warm morning breeze.

A rotten odor, like bad eggs, emanated from what appeared to be a layer of black mud, dry and cracked into irregular shapes. It surrounded the skeletons as if tar had leaked from the bodies and later dried.

Two weeks ago, Red Sky walked away from the Navajo Reservation in northwestern New Mexico to travel southeast to Coyote Gulch. A great battle between Death and the Life Bringer would soon be fought. The Navajo warrior carried a sacred lance carved from oak, with a sharp flint tip, and decorated with a single eagle feather. He was also armed with one knife and a stolen rifle.

Red Sky's grandfather had predicted this day would come. He'd spoken of how their ancient ancestors of Chaco Canyon had battled Death and lost, because the people forgot to honor and give thanks to the Bringer of Life, and were thus abandoned by the Great Spirit. They were wasted by Death in the shape of drought, famine, and disease, but this time, Red Sky's vision showed that Death came as a crawling, hopping, clicking black horde, of flesh-eating little monsters.

He saw it from the feathered back of Life Bringer, who'd taken the shape of an eagle. From above, the wasting appeared as a black shadow, which spread across the land, and nothing lived in the darkness left behind. Death, in the shape of a man, led this black plague. The spirit eagle flew southeast, and there ahead in the living light, Red Sky saw a lone coyote standing at the bottom of a gully, snarling at the approaching black wave and baring its teeth defiantly.

"Death comes to all things," Grandfather had told young Red Sky, "even to the mountains and the seas, but Life Bringer replenishes the land, and the spirits of the dead oversee the proper caretaking, but the Bringer of Life shuns those who forget him, and then the world, becomes Death's wasteland for a time." When Grandfather died, he gave the sacred lance to Red Sky, saying, "This spear came to me in a vision and was made real, and now I pass it on to you, Red Sky."

"But why Grandfather?" He asked. "What must be done?"

"You will know and understand if the wasting comes."

Red Sky told no one of the apocalyptic vision. He had no family because he'd also lost faith in the old ways, much as most Navajos had done everything fell apart. The next day while out for a walk, contemplating the meaning of the experience, an eagle feather drifted down from the sky and he caught it in his hand. He looked up, but there was nothing in the sky except one fluffy cloud. The white blob transformed, as he watched, into the face of a young boy! And then black bugs vomited out of a hole, that opened like a mouth in the cloud before vanishing away.

Red Sky's grandfather's spirit had sent him these signs, and they demanded action. The very next night, Red Sky walked away from the reservation unnoticed. Another would happily take his small share of watery supper slop and tiny living space. He stole a rifle but not a horse. He would walk to Coyote Gulch and hunt his food as he traveled. Red Sky worried not that he'd be hunted down and returned to the reservation. No one cared about one lone Indian on foot, and besides, he'd left a drawing of a circle on the dirt floor in his hut, which symbolized his intention to return.

Red Sky filled his water bag from the creek near the death camp and saw the sun glinting from a shiny object lying on the leaf-shadowed creek bank, sunk into a pile of clothes. Probably just another skeleton, but the reflected sun's glare from the mirror-like object nearly blinded the Navajo, and he approached for a closer inspection. Red Sky laughed when he realized the shiny, polished thing was a flesh and blood human's bald head!

It wasn't a skull, though the cheeks were hollow, and the lips thin, skin clung tenaciously. The prone man wore a long, straggly beard on the other end of his slack face, which pointed into the hot sky, eyes closed; his concave chest rose and ebbed in slow breaths; he lived!

The cadaverous man, unconsciously but instinctually, death-gripped a half-empty whiskey bottle in one hand, which stood upright by his side. The Navajo smelled the stench of the supine man's alcohol-laced breath, kicked him in the ribs several times, and splashed water on the man's face. The drunkard sat up abruptly, then groaned and fell back, clutching his head.

"Yeave me ayoln! Oh God, help me!"

Red Sky answered, saying, "He did; you alone still live." Red Sky observed scabs and weeping sores dotting the man's exposed flesh in a few places, but not many, and he wondered why the man was spared by the black, hopping death.

"Huh? Live where? What the Hell ya tawkin 'bout injun man?"

Red Sky clucked his tongue, click-click-click, and the rumpled man turned his head to one side and vomited. He wiped his mouth with a filthy sleeve and asked, "Are they gone? They didn't eat me like they did the other fellas, but they jumped and crawled all over and bit me some. It was awful, and I just needed to drink and forget about it."

The old Navajo said, "I am called Red Sky, and you are coming with me if I have to drag you along. You saw, yet you live. The Life Bringer has seen. He watches and waits for the people to battle Death, and you somehow are part of the plan. Get washed up!" The stout sixty-eight-year-old Indian picked up the soiled, stinking drunk by the collar and the seat of his pants and threw him into the creek.

The drunk sputtered and thrashed around for a while until he realized that he held one hand high above the water, and it gripped a bottle! He stood knee-deep in the creek, greeted the bottle warmly with a kiss, and gulped down the contents in six or seven swallows. He tossed the empty away, smacked his lips, and told Red Sky, "Okay, I'm ready. Where we goin'agin?"

Red Sky grabbed his rifle and the sacred lance and turned south. "Let us go, Ready. We must warn the people of Coyote Gulch, for Death leads the wasting blackness to them."

"You don't talk like no regular injun, and my name is Charlie, not Ready. Wait; you mean we're walking?"

Chapter 3

Sarah Peale's clientele had grown a lot in the five weeks she'd taken up residence in Coyote Gulch and started taking in laundry, especially amongst the bachelors in town. She sewed and repaired clothes also, and suddenly there was a great need for her personal services. Sarah rented a room in Mrs. Cally's Boarding House and used the widow's well water and property to wash clothes. Sarah offered to pay the widow Cally ten percent of her earnings, but she declined, saying, "Sarah, you are doing me a favor, I just can't do it all myself anymore." The widow cooked, cleaned, and washed clothes for her several boarders until Sarah came along.

No one Sarah had met so far remembered her as a little six-year-old girl named April, who stopped in town once in a while with her peddler father, Jethro Kayne, fifteen years ago. When she'd left her father and his savage pets behind last month, after the slaughter of Foster and his horse, She'd stopped at the miner's camp, where she ate a meal before moving on to Coyote Gulch. While at the camp, April told the miners about the massacre of Foster and his horse, and she worried aloud that her pa would visit his destruction on them as well.

Every miner knew the other, and as one, they vowed to avenge Foster's death and strike Kayne first thing in the morning after a good night's sleep. They hardly knew how to interpret her wild story, because none had ever heard of man-eating crickets.

After her supper, April departed from camp on a donated mule. Though it would soon be dark, she rode away, heavily laden with water and foodstuffs, for fear that her father would follow closely on her trail. She wanted more distance before bedding down. Rule the world, indeed! Ha!

She avoided the lonely ruts called a road and cut across the country, making a beeline for Coyote Gulch. She remembered it as a friendly, thriving town from her few visits as a child with her pa when he stopped to sell his wares.

April walked all night through the warm desert night, using the moonlight to navigate the rocky ridges and steep ravines that crossed her path. However, she employed the hours wisely, constructing an elaborate new history to permanently disassociate herself from her father. She'd be tainted by his sins. Her new identity was Sarah Peale, sole daughter of parents massacred by Indians when she was a youngster, and she'd been raised by an Aunt Mabel Harding in San Francisco ever since. She had reached adulthood and finally struck out on her own to make a happy life. That's the story.

Always emotionally unstable, her pa had finally lost control and gone mad. Pa, always taciturn and hot-tempered, harbored an inferiority complex and was fiercely envious of the successes of 'his betters', as he sarcastically called the well-to-do. His boorish rudeness and outright cruelty made him unwelcome in many of the towns he visited, and he was always altering his routes to find new victims. To Jethro Kayne, even the unintended slight was interpreted as a major slur deserving of retribution, and he knew how to nourish a grudge and keep it alive.

Her pa had grown more frustrated with his lot and thus more belligerent and argumentative over the past year or two, but April never conceived that he could behave so viciously. Sure, he'd smacked her and Ma around sometimes, but he usually only became excessively violent after drinking excessively. April never forgave her mother for running off, leaving her to Jethro Kayne's mercies. She'd do like Ma now and disappear; she couldn't be associated with him.

Sarah Peale knew that her long fine, blonde hair, green eyes, and plump lips, perched on an angelic oval face, would draw men like flies, and she needed one. A man, not flies. Sarah wasn't seriously interested in any of the town's bachelors, but she wished to wed as soon as socially permissible because she was pregnant. It was mercenary, but Sarah needed a connected husband with money, resources, and standing in the town for her future child's welfare. Her happiness was not important, Sarah was used to deprivation, but she swore her child would not experience such travails as she'd endured if it could be helped.

Sarah sat at a tiny desk in her room at Mrs. Cally's Boarding House and read through her list of available suitors again and the notes she'd made about each of them. She crossed one name off the list each time she read through it and whittled it down to three. Sarah had used all her feminine wiles to learn all she could about the eligible men in town while not seeming the least bit interested in any of them.

Sarah wasn't well socialized, having spent fifteen of her twenty-one years isolated, but she'd learned certain skills from her dad, who used wise backwoods psychology to peddle his wares by asking unrelated questions. He'd tell her, "Whenever something has gotta be said or done, ya gotta make folks believe that it was their idea in the first place to say and do it; see?

"That way, they can pat themselves on the back and convince themselves it's fer the best, whilst it all goes to Hell in a handbasket."

Sarah's three marriage prospects were as dissimilar as a frog from a bird from a horse. On one end of a rather wide spectrum dwelt barkeep Terrence Shultz's twenty-two-year-old daredevil son Casey, who loved horses and guns more than people and wouldn't live to be twenty-three, according to the widow Cally. Living and breathing on the spectrum's opposite end was twenty-seven-year-old, blonde-haired, blue-eyed Vernon Cooper, the spoiled, lazy, no-account son of Best Bank manager Bertram Cooper. Verne's smooth, handsome face was a girl's dream come true if she liked a pretty frame around a blank canvas.

In the middle of these extremes, rode a cowboy from the Carson Ranch, who made the twenty-four-mile trip into town once per week for supplies. Named Joe McMartin, age twenty-three, lithe and lanky, with black hair and a bushy mustache, he seemed a real man's man. She didn't know much about his temperament, but barkeep Shultz of the Cactus Juice Saloon said he never caused any trouble. However, Sarah didn't think she was a good fit for the ranching life.

Sarah hadn't yet told the prominent citizen, Dr. Ford, about her pregnancy. She wanted to delay judgment and scrutiny until after she was married and more respectable. Sarah scratched out all three names, put her head in her hands, and cried.

The next morning, Sarah awoke to find blood on her bedsheets, which scared her for the child growing inside. After breakfast, she walked into town, determined to see Dr. Ford, who occupied an office on Main Street across from Sheriff Sears' one-cell jailhouse.

Sarah wore a broad-brimmed straw hat to dull the sun's bright heat and walked atop the boardwalk, saying polite hellos to all she met. Before she reached her goal, a frenzied group of shouting cowboys rode up in a cloud of dust, their mounts frothing and heaving with sweat and exhaustion. The horses had been brutally used, and many townspeople ran to see what was the fright; Indians? However, the four cowboys were incomprehensible in their frantic gibberish, and they talked one over the other, and the doctor and sheriff were called to attend to their shock and hear their tales. Their horses bore the brand of Carson Ranch.

Sarah saw Joe McMartin in the group of four, and she clearly heard him yell, "They're all dead! Eaten alive, we barely escaped. Locusts, black man-eating locusts!"

No!

Oh my God, no!

Chapter 4

A few days earlier, on one of the last blisteringly hot dog days of late July, Earl Gage drifted onto the dead, silent streets of Bleaker Town in northwestern New Mexico Territory. Two months earlier, when Earl Gage stopped in here, busy feet clapped the wooden walkways, horses and mules lined its dusty Main Street, and children played in the alleyways, but now his boots knocked on wood alone, and his pony and pack mule stood by themselves in the dirt road, amongst the bones and skeletons of the town's previous residents. A crumbly black substance, like dry tar, overspread the ground and everything lying on it, cracking beneath his boots with each step, like the sound of crunching ice.

The deathly stillness chilled Gage, even in the building noontime heat. There were no raucous ravens, for all the bones were stripped clean of flesh, but millions of buzzing flies crawled all over the crackled black clay or whatever it was that coated the clothes, bodies, boardwalks, and road. It wasn't dry blood; it stunk of sulfur.

Fully and partially clothed, white human skeletons of all sizes lay in front of shops and homes, and reins, saddles, shotguns, and revolvers accompanied their previous owners, including horses, mules, dogs, and people. Blood stained the surface beneath the bones, and spent shotgun shells by the hundreds littered the ground. The putrid blackness blanketed it all as if inky sewage had rained down from the sky.

Earl doffed his hat, wiped his brow with a red check bandanna, and slapped the dust from his clothes, abusing said hat. He couldn't comprehend the enormity of, nor imagine, the forces able to eradicate an entire town and leave no traces behind but for a black sludge. No fire or plague, no attack by renegade Indians or gang of outlaws, could manage such a massacre; at least eighty people had lived here! Gage didn't figure that any animals or maggots reaped the bounty of dead meat either, but then again, he had no way of telling when this disaster struck. The desolation seemed so complete that Earl imagined the ghosts had even fled the scene.

Just to be safe however, he held a Colt revolver in his right shooting hand and scanned from side to side as he strode towards the Little Bank of Bleaker. Earl entered the business through a shattered window. It seemed odd that the shards of glass lay outside as if someone had broken out of the bank rather than into it.

Gage found the cash drawers empty and the safe locked. "Dammit!' he swore out loud, adding, "The bank musta been closed when catastrophe struck from out of the sky, like vanishing invaders from outer space or something. This is crazy." The sound of his voice seemed somehow comforting to Earl in the stillness of death. A newspaper on a desk bore a date from three days ago!

"Impossible!"

Earl holstered his gun, exited the bank, and stepped over and around the remains of the men, women, and even several children of dead Bleaker Town. Many human skeletons lay in front of doors outside homes or shops, with shotguns by their sides. Shell casings numbered in the hundreds, and they speckled the ground red, but the people had not been shooting at each other. Most of the fallen faced northwest, towards the copper miner's camp.

Bleaker's reputation was unpalatable, it harbored robbers, rustlers, and ruffians, and Earl knew several of them, though they were unrecognizable now. It seemed unfathomable to Earl how such a mean bunch of reckless brutes were completely wiped out without claiming a single kill. He supposed that some of the skeletons might belong to the enemy, but that still left the question of; who or what ate them.

Earl Gage, a twenty-eight-year-old ne'er do well drifter, sometimes horse thief, and cattle poacher, needed refreshment. Trying to swallow the horror of the massacre required the further absorption of booze as a sort of chaser. He clomped, spurs jingling, into the Paradise Saloon and Dance Hall. If he hadn't been so thirsty and traumatized, he would have turned around and vacated the premises. It was an abattoir; blood stained the floor in great quantities amid several dozens of clothed human skeletons, adults of both sexes.

Earl estimated that half the town had taken refuge in the Bleaker Alamo saloon to make their last stand. Obvious signs of the terrific battle, preserved as if for a museum exhibit, were evident. Tables and chairs lay broken in puddles of shattered glass bottles and playing cards. The vacant skulls of the deceased wore fear-like skin, mouths agape, as if screaming as they died. Earl found that most men had emptied the six-shooters, which now lay on the blackened floor next to them.

Gage noted that the walls on either side of the entry and the door were pockmarked with holes but only near ground level. The floor was drilled, too, so the defenders were obviously shooting downwards. They'd either missed every shot, or the bloodless victim's bodies were removed from the battleground, along with every other trace of ever having been here. Neither option made sense, nothing here did. Gage wasn't superstitious. He didn't much believe in anything worthwhile, but this town was cursed while alive and now, more diabolically so, in death.

Earl saw a path through the bones, guns, clothes, and debris; somebody alive had kicked their way to the whiskey behind the bar after the slaughter. Earl pulled his gun again, just in case the man was still around, and he followed the ready-made path to the liquid treasures behind the bar. A broken clock on the floor read 12:37, which he assumed meant after midnight. The shiftless drifter selected a bottle of whiskey from a shelf, found a smudged glass, and poured a drink. The first gulp steadied some of his nerves, so he drank a second glass to quell the rest of them. Earl knew he'd have to rescue a couple of the orphaned cases of whiskey from the saloon's storeroom; he didn't have the heart to abandon them to obscurity in this God-forsaken graveyard.

On his way to the backroom, Earl passed by the lone upright table in the joint. He stopped, when he noticed a half-full glass of whiskey. Something black floated on top. A dead, bug that wasn't a bug. He wrapped it up in his bandanna and slipped it into a vest pocket.

Earl Gage strapped one case of the precious juice to each side of his mule and then went shopping. He emptied cash drawers in every shop and store he entered and even ransacked the pockets of the dead. Earl also packed an empty flour sack with beans, and coffee, from the general store: donned a new hat and new boots from a clothier: and strapped a new saddle onto his horse. His shopping bonanza was completed with a new shotgun, revolver, and some shells and bullets. Then he made ready to leave this haunted ghost town. No way was he spending a night here!

Gage wasn't happy about departing this necropolis right yet; he remained mystified, which troubled him. He wasn't going to hang around, though. It seemed as if the ghosts were trying to tell him something, but he didn't feel like listening. He'd found no reason or cause for this implausible catastrophe, but two nights ago, something laid waste to Bleaker and vanished!

Earl didn't want to run into whatever it was; he wanted out. He lifted one foot into a new stirrup and stopped suddenly as if turned to stone. Someone lived! A voice and vision in his mind; a boy! It made no sense, so he acted rather than think on it.

Earl shot a bullet into the air and shouted, "Hallo!" An echo returned his salute, but that's all.

He walked down Main Street, over and around the skeletons, but silence reigned, except inside his head, where he now heard a muffled knocking. The rapping repeated over and over, and then Earl realized it wasn't just in his head! He followed the signal through the broken door of the Bleaker undertaker's open storefront, where he discovered that the tapping came from a casket on a workbench. A heavy anvil pinned the lid shut.

Earl put the anvil to one side and slid the wooden cover off the coffin. He wasn't completely surprised to see a barely conscious boy with bloody knuckles lying inside the stuffy, stinky box. Gage nearly vomited at the smell as he lifted the limp child, soiled in piss and shit, out of the coffin and set the boy on the front counter.

The child, no older than ten, choked and coughed but lay

with eyes closed, his little chest dipped and rose with each shallow, rhythmic breath. "Stay there, and don't move," Earl said firmly. "I'll be right back; going to get you some water." That was stupid, the poor kid can't move.

Gage trotted back to his horse, grabbed his canteen and a few pieces of buffalo jerky, then went shopping at the clothier's again and found the boy some clothes. Finally, he returned to the undertaker's. The kid lay on the floor! In the front lobby, he hugged the two skeletons there, tears fell from his eyes, and the boy choked in grief.

Earl lost his four siblings to measles, his father to the Civil War, and his mother succumbed to alcohol a few years later, so he kinda knew how the kid felt. He said, "Hey, drink some water and eat some jerky, and then I'm tossin' you into the horse trough outside the livery and throwing those nasty clothes away. See here?" Earl held up boots, a belt, trousers, a shirt, and a hat. Everything's gonna be alright now." His Ma, damn her soul, had always told Earl, "It's okay to lie for a good cause." He set the clothes down at the front desk.

The kid said no word, but he sat cross-legged on the floor, wiped the tears from his cheeks with swollen fists, and drank from the proffered canteen. Earl pulled it away after a few gulps and said, "Whoa, boy, not too fast. Chew some jerky. My name is Earl, by the way. You was most dead, you know."

The boy pointed to his mouth and shook his head before biting a chunk from the strip of jerked meat. The drifter asked the kid, "Are you mute?" The boy nodded yes but used a finger to mime printing words. Earl let the kid drink some more water.

Gage found a few pencils and some unused invoices inside a front desk drawer, and he handed them to the kid. "You write while I fill that horse trough from the water pump across the way there. You stink, boy." The kid smiled crookedly, placed the paper on the floor, lay down on his belly next to the skeletons of his parents, and started to write.

Earl led his horse and mule to the water trough first and let them drink while he grabbed some soap and a bath towel from the general store. Once his animals were watered, he let them feed from the livery's oats and hay. Then he levered a hand pump next to the trough until half full of fresh water. Once all was ready, Earl returned to the undertaker's, a man who'd never be taken under himself.

The boy handed his rescuer one sheet of paper filled with large, neatly printed words. Earl took it and led the kid, who walked stiffly, to the makeshift bathtub. While the mute child stripped and washed, Earl Gage read the epistle. 'Thank you for helpen me, Erl. my name is Danny,' it began, and continued, 'The black snapping bugs ate everyone, so my folks put me in the coffin, but I died anyway, but an indin came, and the sky got all red. And a spirit person sent me back becus deth is alive and I haf to help stop it. Its inside my hed like in a pitcher book.'

Earl heard of people hallucinating when they were close to death or got snake bit. He didn't put much stock in the traumatized child's tale but couldn't imagine the fear and horror the boy suffered either. At least he knew the kid's name.

Once Danny was clean and dressed in new duds, Earl took him into Grubby's Cookhouse and served up a meal of cold beans, stale bread, and hot coffee for the two of them. Danny cleaned his plate. Earl knew they should skedaddle, but he wanted to show Danny something first. He extracted the bandanna from his vest pocket and unwrapped the drowned bug he'd found in the whiskey glass.

The boy jumped up from his chair, eyes bugged out in recognition and fear. He nodded his head vigorously, clucked his tongue loudly, over and over, and put his hands over his ears. Earl tried to imagine thousands of these black locust things descending on a village and wiping out every living creature and failed at it. He shivered and stood, too, it'd be dark in a couple of hours, and he wanted to put some distance between himself and this carnage. He asked:

"Danny, what happened to all the dead ones?"

Danny shrugged his thin shoulders. How would he know he was inside of a dark box! How could he know? He mimicked eating, and Earl assumed him correct. If the nasty little critters could consume a few tons of human and animal flesh in less than two days, they could also clean up after their fatalities. The inky black spots that coated most every surface were poop! Scary!

It was time to leave. He said, "We can't 'stay here, Danny, and my lifestyle doesn't allow fer no kids, so I'll be takin' you down to Coyote Gulch. Some good folks there will sure 'nuf take you in. You ain't got many options, kid." Earl didn't like it, but he couldn't just leave the kid there.

Danny didn't like it, but he knew the man was right. Of course, Earl wanted to be rid of him because Danny was a burden; his father always said so. Well, he wouldn't burden his pa anymore now.

Chapter 5

General Jethro Kayne stood on top of the world; life had never been so good! He sat in Mr. Royce Carson's favorite armchair, smoking one of the man's cigars and drinking his brandy, dirty boots propped on a Turkish ottoman. No one lived to protest. The Carson's private home was spacious and luxurious, and there were enough horses and cows close by, that his growing army of clickets could dine and multiply for many days, before moving on to Coyote Gulch in force. It was too bad those four cowboys rode off instead of choosing to fight and die, because they'd warn the townsfolk. However, that fact didn't worry Jeth; nothing could stop him now.

Qweequa kept Jeth apprised of the general population numbers of her race. According to her reports, total fatalities from the battle at the mining camp and Bleaker Town exceeded twenty-four thousand, and they'd lost another nine hundred, fighting the Carson's and their cowboys last night.

Less than ten thousand remained, but most were females, and the cattle ranch provided a good opportunity for recuperating their numbers. In five days' time, there'd be half a million or more! That's when, on one dark night, they'd hit Coyote Gulch, where lots of food walked about on two and four legs.

The clickets died easily, one at a time beneath the boot heels of men or blown away by the dozens from a shotgun blast, but nothing could hold back the chomping, clicking hordes, any more than one could hold back the ocean tides with the sand castle. The general smiled and closed his eyes to review his successes so far.

After his worthless daughter ran off, just like her damn mother had, Jethro planned a raid on the mining camp six miles distant. He speculated that his daughter would spend the night there, and it would be too bad if his troops accidentally ate the little brat. Qweequa approved. Before delighting in Foster's agonizing death, Jethro had never known the joys of bloodthirsty sadism, and now, like whiskey to a drunk, he wanted more, lots more.

General Kayne started at midnight, leading his army southeast toward the miner's camp. Qweequa didn't travel overland with her kind but rode Jethro. She clung to his collar or crawled into his beard and even under his hat because the Progenitor's life was of utmost importance. Unnecessary risks were avoided.

Kayne walked slowly, at a two-mile-per-hour pace, and five thousand clickets crawled and hopped along behind their general, eating every plant and animal in their path as they marched, leaving a black trail of gooey droppings behind. After three hours of travel, they came to a creek heavily bordered by willows and cottonwood trees. The miners camped on the opposite bank.

All was quiet there, for now.

Fortunately, the miners often crossed the stream, and their

crude bridge led the clickets over to the camp. General Kayne remained on the far side to relish the assault from a safe distance, with Qweequa perched on his shoulder. He wouldn't see any of the carnage through the heavy brush on both sides of the creek, but he could listen and imagine. He had no weapon, but the miners did, and who knew what kind of crazy shooting they'd do to defend themselves.

Jeth found out soon enough after the attack commenced. Shotgun pellets and bullets flew, and screaming men ran amok, with several crashing through the brush and jumping into the shallow creek water. The choking, screaming, and cursing sounds of the dying were almost too horrific for him to appreciate, but he managed with a smile.

Afterward, Jethro Kayne and his loyal clicket troops occupied the camp for two days before consuming all the meat. Jeth ate the miner's grub, they wouldn't miss it. Bleaker Town was the general's next target. In his peddling days, he'd been forced to make a wide detour of the Bleaker den of thieves because the low-down rotten cowards always harassed him and made him the butt of their jokes and pranks. They'd pay for mistreating him.

Genera Kayne led from the front once more, with his force having increased to nearly forty thousand soldiers. The hordes departed the miner's camp in the late afternoon and mowed a fourteen-mile long patch, fifty yards wide, to Bleaker Town in less than seven hours. They'd made good time, so it was only half past midnight and earlier than he wanted to attack; too many people still walked the dusty streets. Qweequa assured her general that all would be well, and Kayne acquiesced and ordered the attack.

The living carpet of crawling, hopping, black death silently approached Bleaker's outskirts, massed together and keeping to the darkest shadows where possible. Kayne saw several groups of faint human shapes hanging around outside the noisy saloon, silhouetted in the light emitting from its windows and open door, but much of Bleaker was dark and abed.

Thousands of creeping clickets silently entered sleepy homes through open windows and down chimney pipes and caught families in their beds. Unfortunately, Jethro couldn't enjoy the festivities in person and had to imagine the horrors of waking up to being eaten alive. The first shrieks of terror to shatter the night grabbed the attention of the carousers near the saloon, and they paused to learn the source. The second series of blood-chilling screams temporarily paralyzed those toughest, meanest men in the west, as they often boasted of being. Instinctively they drew their handy guns but hesitated to act, not seeing nor hearing an enemy.

Suddenly bedlam erupted in a cacophony of gunshots, screams, and yells, and the mind-shattering, clicking thunder of chomping jaws! The town came alive, only to die. It required less than three hours for the Kwarliks to annihilate the four score humans of Bleaker Town. Clicket losses were heavy, but the surviving thousands happily chomped and gobbled the remains of the dead, including their own fatalities. The victory was won too swiftly, and the screams ended too abruptly for Kayne's liking, but he consoled himself with a few glasses of whiskey from the Paradise Saloon. He celebrated the happy victory, especially because Qweequa was satisfied with his leadership and the appropriate choice of targets.

Amazingly, the surviving clickets and their offspring scoured the bones of the fallen within forty-eight hours, paring every shred of flesh and consuming every organ, leaving nothing but bare bones blackened by their droppings and then General Kayne and many thousands of clickets, made for the meaty Carson Ranch, nine miles to the East. He had no beef with the folks there, but an army can't march on an empty stomach, and it promised a huge return for a small risk. So far, he'd run a beautifully successful campaign, and Jeth felt confident of future successes. Ruling the world was gonna be great.

Chapter 6

Lightning, thunder and gusting winds announced Earl's and Danny's early morning entrance into Coyote Gulch, but nobody noticed the two strangers ride into town. Main Street loudly overflowed with excited men and women, and the commotion all centered around four cowboys and their lathered horses. Earl had no more than secured his own horse and mule to a rail and helped Danny to ground when the sky opened up, and water cascaded down on Coyote Gulch as if heaven's dam had failed.

Resembling a coordinated drill, the four cowboys spontaneously ran for shelter in separate directions, and curious citizens followed closely behind each. The largest group made for the Cactus Juice Saloon, while smaller teams invaded the Coyote Den Hotel, Mrs. Cally's Boarding House, and Sam's Restaurant across the street.

Earl randomly followed the particular crew that fled the downpour for the comforts of Mrs. Cally's boarding house, where Joe McMartin found refuge, along with Dr. Ford, Sarah Peale and nine or ten others. Danny stuck to his rescuer like glue, and they assembled together in the widow's dining room.

Matronly, Mrs. Cally, whose seven children busily populated California with dozens of children of their own, fussed about boiling coffee and making her guests feel welcome. She always brightened faces with kind words and a genuine, rosy-cheeked smile. Eight people sat around her long dining table, and the seven left standing crowded closely around it. Earl and Danny kept to a corner, however, remaining unobtrusive for the moment. Sarah helped the widow in her several tasks, cutting bread and cheese, but she listened carefully as Joe told his harrowing tale.

Joe commanded the room from the head of the table, and once all were settled and quiet, he related his gory story, an oral drama repeated, with different exaggerations, in three other venues as well. "Me and the boys," he said, "was sacked out and sound asleep in the bunkhouse when a huge commotion erupted outside. All kinds of racket, a loud clip-cloppin' noise like ya never heard before; sorta like a thousand trains clicking over the tracks all at once. So we all threw on trousers and boots and run outside, and the noise, like a million saws goin', was getting louder all the time, like whatever it was, was comin' closer. All the Carson's, 'cludin' the three kids, was all out in front of the house, closer to the clicking things than we all was, and facing away from us, in the direction the infernal racket was a 'comin' from."

Joe paused to take a breath, but his listeners were breathless. After a moment while waiting for dramatic tension to build, Joe continued his harrowing tale. "'Twas darker than inside a cow, and we couldn't see nothin' just hear somethin'. We were looking out to the horizon, not at our feet, and then afore ya know it, the Carson's was submerged 'neath a black tide-like. These jillions and gazillions of black locusts was hoppin' and crawlin' and bitin' 'em! Some of the boys run to help 'em, but it wasn't no use, and me, Hal, Eddie, and Dirk ran for it, weren't no sense in tryin' to fight 'em off. Look at my neck and arms!" Several weeping sores, like gouges where a chunk of flesh had been carved out, were a testament to his tale.

Dr. Ford said, "I want to. You and the others need to come into my office. Tonight! Like right now!"

Joe said, "Sure, Doc. But lemme finish. See, my pals and I jumped on the closest horses and rode like Indians without saddles and reins. We stopped a safe distance away but couldn't see nothin' back at the house, but the shrieks and screams we heard, scared the Hell out of us, and we rode like all the devil's hounds was chasin' us! They was gitten eatin' alive! I swear it! They's the devil's children, I tell ya!"

Mrs. Cally casually poured coffee for the four sitting at the table who wanted a cup as if she heard such tales every day. No matter how dreadful the stories, they always went down better with refreshments. Sarah followed behind the widow and set out two platters, brimming with cheese, and bread, and a slew of cookies. She asked, "What did it?" She knew, however, and wished she didn't.

Earl stepped forward and spoke up now, answering her question. He said, "If it's what hit Bleaker Town, then all of the Carson Ranch people are dead and being stripped of flesh; eaten." Gasps, cries and protests erupted around the table, and a dozen glares bored into the tall stranger with the bushy mustache. Earl reached over Joe's shoulder, and set his red check bandanna on the table in front of the cowboy. No one spoke, and then, slowly, dramatically, Earl unfolded his kerchief, to display the dead, black bug thing he'd found floating in whiskey.

Joe scooted his chair back, and yelped, "That's them!" Where'd you get it?" Dr. Ford, who sat across from Joe said, "What's this all about, and who are you? I ain't seen you around before."

Earl dissimulated. He said, "The boy with me is Danny, he's mute. I pulled him out of a coffin in Bleaker, where his folks hid him from these creatures. He's the only survivor. And this was the only bug, and I bet you won't find any at the Carson Ranch either. Apparently they eat their dead, along with everything else; because when they leave a place, nothing but bones remains." Earl saw no need to reveal his identity, he wasn't on any wanted posters, but his name might mean something unhappy to ranchers and lawmen in the area. Anonymity was safest.

Shouts of denial erupted from all sides, and Earl was vehemently accused of lying by some. Dr. Ford cleared his throat, and after the initial clamor quelled, he asked the stranger. "You are standing here telling me, telling all of us, that these itty bitty critters killed and devoured every last lost soul in Bleaker Town. Is that correct, Sir?"

Earl said, "Precisely correct, and little orphan Danny can confirm it. His parents were victims, you know."

Sarah felt dazed, near fainting. She said, "A whole town; gone." The realization felt like a knife in her chest, piercing her heart. Her father? And his clickets? It just couldn't be! Sarah knew her inner protests were futile, but she couldn't accept the truth yet; denial would suffice for now.

Mrs. Cally exclaimed, "God rest their rotten villainous souls."

Dr. Franklin Ford pointed at the black bug thing and said, "You mind if I take it? I want to look at it through a magnifier." Earl said, "Take it. You can have it."

The doctor asked while refolding the bug specimen into Earl's bandanna, "Hey, Nameless, if these things eat their own dead, then how did you come by this one?" He carefully placed the item into a vest pocket, where the pocket watch he didn't own was supposed to perch. "It drowned its sorrows in a glass of whiskey,"

Earl replied, not taking the bait.

Sarah felt thankful that nobody knew of her father's involvement in the massacres in Bleaker and on the Carson Ranch, but it was now more important than ever that she protect her identity. Sarah knew in her aching heart that Pa had used the miner's camp as a test run for his clicket army; she just knew.

Earl said, "Doc, I think a man is involved, but maybe not." Sarah choked on her coffee and made an excuse to leave; "I don't feel very good. It's all too much," and she ran outside to vomit. She never noticed the wet streets or the sun peeking through thinning clouds. A blue sky was visible to the north. and a short, faint rainbow arc hung like a halo in the western sky. She sat on a bench on the widow's porch, nauseous, face in her hands, listening to the eaves dripping like tears.

Dr. Ford was concerned for her, but he had to know. He said, "Spit it out, stranger! What man?" Earl shrugged his shoulders and turned his back to answer as if embarrassed by his words. He said, "Someone went into the Bleaker saloon after the slaughter, but before I got there. The person or persons kicked open a path through the bones, and guns and chairs, from the door to behind the bar, and drank some whiskey, I assume, to celebrate the victory."

Dr. Ford huffed, stood up and said, "Come on, Joe, time to round up your wounded and come down to my office, past the barbershop." He abruptly departed. Joe grinned, said to Mrs. Cally, "Thanks fer the coffee Ma'am," and followed the doctor outside.

The pair found Sarah sitting on the bench, and they expressed their concern. She assured them, "I'll be okay, but all those people are dead and maybe eaten alive; it's too much to contemplate."

The doctor nodded and said, "Then don't think on it, Miss Peale. I can give you something for nausea and nerves, too, if you like."

"Thank you," she said, "but I'm alright now." However, Sarah did wish to consult him about her pregnancy at some point, but now wasn't the right time; the doctor had patients to attend to.

Dr. Ford told Joe, "I'll be waiting in my office," and he walked off in that direction. Joe went into the saloon to collect a wounded comrade, leaving Sarah to herself.

With Joe gone, his audience soon filed out, leaving Earl and Danny with Mrs. Cally. Earl paid the widow for a room, but he would be forced to share a bed with Danny because that's all she had available. Earl explained to Danny and the widow that a man needed a drink of whiskey at times like this, and with their assent, that's what he went out to do. Mrs. Cally made the boy pancakes for breakfast.

Earl found Sarah sitting outside the door, and he politely doffed his hat and said, “I ain’t gonna ‘pologize fer saying what needed sayin’, if that’s what’s ailin’ you Ma’am. People need to be warned, ‘cuz them creatures could be coming here next. That ranch ain’t so far away.”

Sarah wasn’t in the mood for conversation and retorted harshly, “Yeah, you’re just going to move on anyway and leave the boy here to die with the rest of us then. Thanks for your concern.” She knew a drifter when she saw one; they always swaggered about like they owned the world, and nothing good or lasting mattered to them. They were no better than predators.

Earl didn’t expect such sass from a pretty young woman, and he hotly responded. “Stay here and die then! It don’t make no never mind to me. I’m getting me a drink first, but I’ll tell you something else; I seen you in there, tryin’ not to act like you already knew what I was talkin’ ‘bout.”

Sarah averted her eyes, but not before Earl saw her startled flinch, which confirmed his suspicions. Somehow, she had a connection to all this. Everybody had an angle, and no one could be trusted. She was hiding something, just like he was. An evening breeze toyed with the coils in her long blonde hair when he said, “I’m bettin’ you ain’t been in town too long.” Then, he turned his back and went to see his animals.

Gage led his horse and mule down Main Street to Potter's Livery and Stables and got them into stalls. Earl negotiated with the head stable hand, and they agreed on a bribe of two bottles as the price of hiding the two cases of whiskey he'd purloined from Bleaker. The conspirators covered the booze under a pile of hay in an empty stall, and then Gage headed back up Main Street to the Cactus Juice Saloon for a few drinks. When he passed the boarding house, he noticed that the sassy woman no longer sat outside. He wondered what her story was; she seemed as secretive as he.

Earl thought about getting out of town immediately, but after a few drinks, he changed his mind. He hung out in the saloon most all day and night, breaking only for a forgettable diner meal, playing cards and gambling away the money he picked up in Bleaker Town. It irked Earl to realize at some point that he was purposely avoiding Danny, actually hiding from the kid.

Sheriff Sears delayed leading a posse out to the Carson Ranch until tomorrow because he wanted the four survivors of the massacre to ride along, and Doc Ford said they needed time to heal from outer and inner wounds first.

Chapter 7

Earl Gage's plan to sneak out of town early next morning was kicked to dust by seven mounted men; the four cowboys from Carson Ranch, barkeep Shultz, Dr. Ford and Sheriff Sears. He'd overslept. The Sheriff and the doctor rode up close to Earl when they spotted him leading his saddled horse out of the livery stables. He hadn't yet packed his mule. Earl eyed the approaching pair warily and stopped to hear what they had to say.

Thirty-nine-year-old Sheriff Archibald Sears, six feet three inches tall, looked as big and brawny as a bear, sitting on top of his chestnut mare. He looked down at Earl and said, "Hey, Mister, I appreciate you riding out to Carson Ranch with us. That's mighty civic-minded for a stranger 'round these parts. Frank here, I mean Doc Ford, says you know something about what happened out there." His tone and manner brooked no opposition, which Earl Gage naturally opposed.

He introduced himself as Earl Taylor and said, "Actually, Sheriff, I ain't too civil or civic neither, and I wasn't planning on…." Earl felt a tugging on his shirt tail and turned to see Danny behind him. The dumb boy's brown eyes peered up at the drifter, pleading as words could never manage, but they stabbed Earl with piercing accusations of betrayal. It hurt too, and that made no sense to Gage. He thought he'd rid himself of emotional attachments and tossed them off like a bronco throws a rider. No kid was gonna rein him in.

Dammit! Somehow the orphan got attached to him. It was downright embarrassing that the kid saw right through him and knew him for a phony, no-account loser bum. Gage thought he hid it better; he'd have to work on that. Meantime's; he was outta here! He'd forewarned Danny that he was no good and not gonna be his babysitter.

Sarah Peale came out onto the boarding house porch now, scanning this way and that, looking for Danny since his breakfast was on. She called for him once, and Joe perked up when he heard her voice. Earl responded quickly and hollered back, "Over here!" The woman would get the kid off his back.

Sarah walked out barefoot onto Main Street, amongst the milling horses and people, to retrieve the orphaned child. "Good morning!" she greeted once to cover everybody there, and she took Danny's hand in hers, and Earl mounted his horse.

Earl and the other seven riders doffed their hats and wished Sarah a good morning. Earl thought she looked fetching, with her golden hair tied up, baring her slender neck. Sheriff Sears said,

"We better head out. It's a fifty-mile round trip, and today is gonna be a hot one."

Sarah started to lead Danny out of the street but turned her pretty head up to look at Earl Gage and say,

"I'm Sarah, Sarah Peale, by the way. I'm glad you're staying around for a while."

"Earl, um, Taylor. Nice to meet you, Sarah Peale." She turned and led Danny back across the street to the widow's place, and the drifter thought, "Are you talking to me; a man who never stops moving on?" When people asked him why, Gage always told them, "I ain't running from nothin', I'm runnin' to somethin', but I just ain't learned what it is yet."

He gazed in admiration at the swing of Sarah's hips and her lovely ankles as she walked away, trying to decide if he'd decided on a course of action yet, when Sheriff Sears yelled, "Let's move out!" The lawman wheeled his horse around and set off on the well-traveled road north to the Carson Ranch. Seven other riders fell in behind or alongside, and Earl was surprised to find that he was one of them.

The posse rode the well-traveled Raven Road, which led between the ranch and town and nowhere else. The dusty tracks twisted around rocky outcrops, dipped down into dry washes, and rose upon bare, windblown ridges for almost twenty-five grueling miles. Earl let his horse do the driving, following the others.

He rode passively and let his thoughts run amok. He didn't

know himself suddenly and wondered what he would do if they met an oncoming horde of the little black monsters: gallop away and never look back or fight and die. Earl Gage wasn't a coward, but self-preservation was a priority, and he didn't owe these people anything. He wondered what his motive was in coming along, but introspection left him exhausted, so he put his restless thoughts down for a nap and enjoyed the exhilarating ride. It was a gorgeous morning, with cloudless blue skies and not yet hot.

It was hot when the octet arrived at the property's gates three hours later, without incident, and rode beneath the iron and stone archway, which read, 'Carson Ranch,' bookended with the CR brand. Sheriff Sears and Joe McMartin led Schultz, Ford, Gage and the three former ranch hands toward the main residence of the five Carson s; parents and three adult children. Two passing crows cawed and argued overhead before vanishing into the bright distance.

The eight riders slowed their mounts and guided them respectfully, single file, around the eleven white skeletons that lay in the dirt of the main drive-up. The mass boneyard rattled Dr. Ford more than had his experiences as a medic in the Civil War. Hot air shimmered and rose into the air as if the ghosts of the dead were ascending to Heaven. None of the posses felt invincibly tough at the moment like every self-determining individualist was supposed to do. Instead, they felt shrunken and diminished in the silence of so much death.

Dr. Ford's voice made a few men jump when he asked aloud, "Does anyone else wonder what all this black smelly smudge overlay is about?"

Shultz drawled, "It shor ain't dry blood Doc, but it's thickest around the bodies like it come out of 'em."

Earl said, "Figure it out, Doc. Something that eats and digests continuously also poops non-stop." He had a pet like that once. He'd found a young rabbit and taken it home, and eating and pooping were what it did best until Ma went behind his back and cooked it for supper one night. She didn't tell him until he'd had seconds.

"My God!" the realization shocked the doctor. "Look at it all; tens of thousands of them must be! And their metabolism must burn hotter than a steam engine!"

Sheriff Sears didn't try to understand the enemy, and he confronted it. He announced, "I'm going inside. Joe, you come with me. You know your way around in there. The rest of you, keep your eyes and ears open out here."

"You bet, Sheriff," Joe answered back. The pair dismounted, handed their reins to a companion, drew a gun from their holster, and walked between pillars and up to Carson's white front door. It's too quiet even for a cemetery. Earl thought it smelled wrong; something more than shit and death. Maybe he imagined it, but Earl experienced sudden nausea. He abruptly spoke up and said. "That's a really bad idea, Sheriff. There's a man or men involved, and while them critters might have moved on, they could be lying in wait somewhere too."

"Well, if yer scared, move on," Sears said. The Sheriff didn't bother to knock. He kicked the door in and rushed inside the house, gun barrel first; Joe was hard on his heels. Seconds later, shots were fired! Shouts of alarm, then Joe and the Sheriff blasted out the door like cannons, engulfed in a crawling, biting black cloak of clicking, chomping mini-monsters. A black wave vomited from the door now, and from under the porch, the eaves, and many of the murderous things, turned their three eyes and clicking mandibles in the posse's direction.

Joe and the Sheriff flopped onto the ground, to roll in the dirt like men on fire, slapping and clawing at the clickets in vain. The attackers filled their ears, nostrils, throats and the screaming mouths of the Sheriff and cowboy, and the men suffocated after a minute or two and couldn't notice being consumed. The six mounted men turned to stone, so it seemed, the sudden shock unassimilated, and they were undecided on fight or flight. Their horses knew to flee; they reared and whinnied and shook their heads, urging their dumb riders to action.

A dirty, unshaven man stepped outside the door in the wake of the bug things and laughed maniacally. Dr. Ford and Earl recognized the peddler when he turned in their direction, pointed at them, and faintly heard a yell, his voice competing with thousands of clicking monsters, "You're next!" Great bouts of hysterical laughter erupted from Jethro Kayne after his threat. The six horrified men wheeled their mounts around and put spurs to them, galloping away at top speed, racing for their lives!

In their blind flight, the six panicked men, putting spurs to their mounts, nearly trampled a pair of travelers in their path. The two men were idling under the archway entrance gate to the ranch, listening to the massacre from a safe distance; it was too dark and far away to see anything clearly. The two pedestrians dove for safety to avoid the stampede.

The three ranch hands fled the property for the second time, galloping away south towards Coyote Gulch as quickly as their horses could run. Earl, the doctor and barkeep Terrence Shultz stayed back to chat with the two travelers they'd nearly run over.

The presence of the strangers here and now raised hackles and seemed suspicious indeed to the mounted trio. Where had they come from, and where were they going? The men felt safe out here, a hundred yards from the massacre, and the clicking killers weren't chasing after them. The creatures moved too slowly to be concerned about anyway, and some questions and answers were required of the two loiterers.

A tall Indian and a thin, disheveled man with a bald head shakily rose to their feet. The former brushed the dust and dirt from his deerskin outfit and picked up a beautifully decorated lance, topped by an eagle feather he'd tossed down in his dive for safety. After a roll in the dirt, the latter person noticed only an improvement in his condition. He spoke first in a hoarse, strangled voice, "I sure could use a drink, folks; it's been like a week, and I'm so thirsty. Anyone?"

The Indian had more important things weighing on his heart. He asked the trio, "Where is the mute boy?"

All were surprised by the question. Earl dismounted to confront the Indian about it, while Dr. Ford remained more circumspect and logical. He asked, "Who might you two be, and what is your business here?" Order! Things needed to be done in proper sequence.

Ready knelt, put his hands together in prayer, and pled for mercy and booze. He said, "I'm Charlie, but he calls me Ready; comin' from the miner's camp up north a spell. Whiskey?"

Earl stood toe to toe with the Indian and sized up the strong, tall, ancient Navajo, with lines on his face like geologic layers, telling tales of lost ages: as sturdy as a rock. A cascade of white hair fell about the Indian's broad shoulders. They were the same height, and Earl glared directly into the Navajo's unblinking eyes. Gage asked him, "How could you know of dumb Danny? Were you the men that needed drinks in the Bleaker saloon?" Earl assumed peddler Kayne to be the true culprit behind the mayhem, but this Indian knew something. But how?

Red Sky, head held high, stared straight ahead, right through Earl as if he wasn't standing a yard in front of him. "I am Red Sky, and it's only been half a day, Charlie Ready since you were drunk last." Then he attended to Earl, looked into his eyes and asked, "And what do you know of this gifted child? He sees and hears the things of the spirit world. He died but was revived by the Life Bringer. He is the key to defeating the wasters."

"Hogwash! You were there!" Earl accused. He denied any stories of the supernatural; his Ma had cured him of faith. She said it was a sickness that made folks…" feeble and weak and depending on prayer to help them when you just have to help yourself!" Indians could hardly be trusted any more than white folks, and maybe considerably less so, in his estimation, especially when they started preaching.

Terrence Shultz lost patience with the exchange and said, "You should be on the reservation anyway. What a pair; Red Sky and Ready."

Dr. Ford pondered the circumstances and interrupted the arguments. He said, “Two massacres, two survivors; perhaps there is a commonality, some shared factor that kept them both alive.”

Earl laughed and asked, “Are you kidding, Doc?” Then he turned to find Charlie Ready, who lay in the dirt, napping. Earl nudged him with a boot toe and said, “Hey, Ready, was you hidin’ inna a coffin?

“Nope, I was drunk and wasn’t doin’ no coughin’. I puked once if that counts.”

Ready’s reply reminded Earl of the dead bug he’d found floating in whiskey, but he couldn’t make sense of it.

Red Sky asked, “Can we get a ride into Coyote Gulch? I must meet the boy.” The Navajo dumped his rifle while avoiding flying horse hooves, which lay unseen in the grasses. It would be useless in the battle to come, so he let it lie.

Earl hesitated. The Indian obviously believed what he said, and as unlikely as it may be that Red Sky and Danny had a connection of some kind, this whole damn mess was improbable, so he relented. In fact, he allowed the Indian to double up on his horse while Charlie Ready shared Shultz’s mount. He’d keep a close watch on Red Sky.

Before they rode off together, the doctor remarked to nobody in particular, "The Sheriff left a widow and a pair of fatherless teenage daughters behind. He died a brave man, and I was proud to know him."

Shultz said, "Amen."

The trio of cowboys spread the painful news of Joe McMartin's and Archie Sears' grisly deaths throughout Coyote Gulch before the Sheriff's family learned of the loss. Joe had no relatives in the area. He came from out East, but all who knew him liked the young man. Sorrow consumed the Sheriff's girls, a widow and two daughters, and anxiety gripped the entire community, strangling happiness and throttling smiles.

Grim faces met Dr. Ford and the others an hour later when they rode into town under a light drizzle of windblown rain. A spectacular orange-red sunset glorified the western horizon when the three men left their worn horses to the care of Jonny Potter's Livery and Stables and went their separate ways. The Doc walked to the Sheriff's office, and Shultz entered the Cactus Juice Saloon with Charlie Ready in tow as if attached. Vic Mason managed the drinks tonight, and the saloon's walls bulged with jittery patrons, drinking for solace from their troubles.

The saloon was heavenly to Ready, who immediately stepped up to the bar and ordered a bottle of whiskey. Barkeep Vic didn't think the baldheaded dirtball could afford a drink, and he said, "Most folks pay for their whiskey here, fella." He wondered how many critters nested in the guy's tangled, knotted gray beard.

In less than ten silent seconds, Ready heard and analyzed the sentence and replied, "Wait." He knew he didn't have a penny but diligently shoved his soft hands deep inside his pockets. His face lit up like a full moon after a cloud opened up. He felt lots of money and heard happy jingles! Charlie's fists clutched coins by the dozens in both pockets, and he suddenly remembered robbing the deceased miners' bodies, tents, and shacks.

He plunked two bits on the bar and said, "I'm like most folks, and I pay."

A few hundred paces west, Earl Gage made the mistake of trying to bring Red Sky into Mrs. Cally's Boarding House. She barred his entrance, stating, "I don't allow no heathens 'neath my roof, Mr. Gage." It mattered not because Danny snuck past the widow and dashed out onto the porch to see the Indian of his visions. Until this moment, the Indian with the tall lance only lived in his mind's eyes. Danny's mind picture showed the warrior standing down in a gully, holding a lance high, with a coyote by his side, facing a red sky. It was real now. The boy knew, too, when Red Sky arrived in Coyote Gulch.

Mrs. Cally huffed and shut her door on the three so she didn't have to witness the interactions. Then she went to a nearby window and peeked at them from behind the curtains. Earl remained attentive when Red Sky, holding his lance like a walking stick, squatted down to see eye-to-eye with Danny. The boy pointed to the red bandanna he wore around his pale neck, then up at the heavens, and finally at the Indian. Red Sky laughed delightfully and said, "Yes, Danny, I am Red Sky." Grins.

Relieved, Earl sat down on Mrs. Cally's porch bench and tried to break his wild thoughts. They kicked and bucked like a bronco, and he couldn't tame the rowdy musings. Whatever was happening was too big to make sense of; it was like trying to understand a woman after catching a glimpse of her lovely ankles.

Mrs. Cally cracked her door open and set a bowl of cold rabbit stew on the porch; "Heathens have to eat, too," she muttered before shutting the door. Red Sky accepted and ate every scrap.

Dr. Ford consulted with acting sheriff Deputy Palmer in Sear's old office about calling for an emergency session at the Meeting House, a multi-use facility financed by local banker Bertram Cooper. Parson Murphy held church services there on Sundays and holy days; school lessons were taught there in the winter season, and court cases were conducted at the Meeting House whenever the traveling judge stopped in. All users paid the generous banker a nominal fee for maintenance, except in an urgent crisis.

The new Sheriff heartily agreed with Dr. Ford, and the two men appointed seven o'clock tomorrow night for the assembly. After the doctor left his office, the lawman immediately sat at his desk to jot down some notes regarding defenses and preparations in anticipation of the likely invasion.

Upon leaving the Sheriff's office, Dr. Ford tracked down Bret Armbruster, the editor of the Coyote Howl weekly newspaper. He found the eloquent but slow-witted, longtime editor in his spacious living quarters above the paper's offices. Dr. Ford was newly astonished by Bret's massive, ponderous figure every time they met, for the man seemed rounder on each occasion. He liked to eat while he worked, and he loved working.

They sat on cushioned wooden chairs at a round table, and Bret listened without interruption as the doctor described the massacre of Bleaker Town and the Carson Ranch slaughter. When Doc finished the dramatic tale, a moment of silence ensued.

Obviously, the doctor is overwrought, Bret decided. Bret liked the nosy ol' duffer, but the physician needs a doctor. And he needs to stay out of other people's business too! That's the newspaper's job! The mutt wanted something from him, so Bret asked, "How can I help you, Dr. Ford?"

The doctor adopted a congenial approach that wouldn't overwhelm the editor. He said, "Call me Frank." He followed up with, "The town needs you, Bret," pouring the syrup on thickly. "This trouble needs every citizen to do what they know best.

Print a special edition right now! Tell the stories, and be dramatic. Scare people; this is a question of survival. I truly believe that! Announce a town assembly at the Meeting House tomorrow night at seven. Print handbills and posters, Bret, whatever you can do. If you need help or funds, just ask."

Bret didn't seem to quite grasp the gravity of the situation, but he enthusiastically endorsed the project; it could boost sales. He asked the doctor, "What do you call the creatures? I gotta have a catchy headline, you know." Bret was edgy, like a racehorse or perhaps a hippopotamus at the starting gate, chomping at the bit, ready to run at full speed until he reached the finish line. He was already making a mental list of the food fuel required to complete the job.

Dr. Ford scratched his short gray beard stubble and hemmed and hawed for a moment before saying, "Good question." He pondered aloud, " They make a horrific clicking sound, but clickers don't strike much fear into people, and we need folks to be frightened half to death, 'cuz dead is what we'll all be if we don't stop the...Wasters! That Indian fella, Red Sky, called them

Wasters!"

Editor Armbruster recruited help, and the next morning kids were vehemently hawking the newly printed Coyote Howl. The boys hollered out, "Special Edition! "They followed up by shouting the headline, "Wasters; Their Clicking Will Be the Last Thing You Ever Hear!" Below, the horror story of peddler Jethro Kayne and the ravages of the Wasters is narrated, first describing the decimation of the mining camp, followed by relating the massacre of Bleaker and the Carson Ranch too. 'We are next,' the editor warned, graphically depicting the suffering horror of being eaten alive. At the story's conclusion, the Howl announced tonight's critical gathering in big, bold type. It read, 'All able-bodied men between age twenty and sixty are requested to attend the Meeting House assembly at 7:00 tonight!'

A perceptible fear and apprehension squeezed the citizens of the seven hundred and forty-seven residents of Coyote Gulch until their kindness and patience oozed out, leaving only the gritty crud of selfishness and rudeness remaining inside. Brusque people rushed hither and thither, stocking up on necessities and even hoarding because family comes first. Often folks seemed to be working against each other rather than combining their resources towards their common goal; defeating the wasters.

When seven o'clock still lay ten minutes in the future, seventy-five anxious citizens already sat on the wooden benches or stood against the walls and in the aisles of the Meeting House, and scores more stood outside of the two-story, red brick building as well. They propped open its double entry doors and stood near open windows so that they could listen in. This allowed some hot air to circulate through the hotter building. Ninety percent were men, and the remainder were women, including Sarah, who sat in the front pew. No children were allowed, and Mrs. Cally cared for Danny in her home.

Parson Murphy opened the session with a prayer and spoke from a raised dais as his usual custom on Sundays. Though nearing eighty years of age, his smooth complexion, strong voice, and full head of snowy white hair, which graced his head like a halo, belied it, for he was fit as a fiddle.

He grasped the sides of the podium with bony fingers and preached, "Men and women of Coyote Gulch, let us pray; Lord, hear our fervent prayer. A black death lays waste to your people and infests the land, just as black sins infest our souls. Dear God, we ask in your holy name to protect and preserve us in this time of need. Wash away the blackness of our hearts and lands; cleanse us of our inner and outer demons. Amen."

A strident chorus answered back, "Amen!"

The parson said one last prayer before turning over the podium to Dr. Ford. He said, "Let us pray, too, for the dearly departed souls of Sheriff Sears, Joseph McMartin and all the others lost to these wasters. And let us not forget to keep Widow Sears and her two daughters in our prayers and daily cares. Amen."

A subdued Amen from bowed heads replied.

Dr. Ford ascended the podium and briefly told the tragic tale of yesterday's experience at the Carson Ranch. Following that, he spent the next several minutes describing the anatomy of the wasters. "They are only the size of a big man's thumb," he stated, "but when thousands of them bite you through your clothes and penetrate into your every orifice, well, you can't breathe, and you die."

He went on, saying, "They are black and scaled like a snake and are not insects. They may not even be from Earth. They have three fly-like eyes and four legs, not six. The front feet are barbed and can cling to anything, and their back legs resemble those of a frog, so they hop, crawl, and climb.

"Instead of bearing wings on their sides or backs as insects do, these wasters have tentacles, two on each side. From other signs, it appears that these prehensile appendages are useful in carrying their eggs with them as they migrate to new food sources. I imagine they reproduce by orders of magnitude. A dissection showed me they are eating and breeding machines and little else. They have no lungs, for example, and apparently absorb oxygen directly into their blood through spiracles or pores under their scales.

"Finally," the good doctor said, "their jaws or mandibles open both vertically, like our own, and horizontally as well, and their beaks are built for ripping and tearing out great chunks of flesh. They took down big brawny Sheriff Sears in seconds!" Few comprehended the entirety of the doctor's words, but every last one of them understood that their life was at stake.

"Wait; one more thing," said the doctor, while scratching his whiskers, "Somehow, I can't even speculate how a man leads them to their next crop of flesh. Some of you might even remember Jethro Kayne, the crusty peddler that used to swing through once in a blue moon before we banished him fifteen years ago or thereabouts for beating a prostitute near to death. Well, he was there at the ranch; used to bring his daughter April along on his route too, when she was a toddler, but we don't know if she is in with him now or her whereabouts."

Earl sat up near the podium with Deputy Palmer, the doctor, and the parson. Because he spoke next, he faced the assembly and surreptitiously watched Sarah Peale. She sat on the wooden front bench, not ten feet away. The drifter, naturally suspicious due to his own dishonesty, slyly studied Sarah's reactions while the doctor spoke and concluded that the woman should never play poker. She squirmed in her seat, with plump lips tightly compressed into thin hard lines, and her hands writhed in her lap like a nest of nervous snakes. He couldn't see her downcast eyes.

Sarah withstood the shock of the Doc's revelation without twitching a muscle, and she felt proud that she hadn't reacted because that cowboy drifter was watching her closely; the creep. She battled the fight or flight urges that coursed through her body and mind and played it cool, calm and collected. She could do this, and she had to.

Earl stood at the podium now. He smelled fear in the air like smoke, stoking the anxiety further when he spoke. "Everyone in Bleaker died, fighting for their lives. None survived, except the boy Danny, who claimed he died. In no more than two days, sixty hours at most, following the slaughter, every bone and skeleton was picked clean, the meat stripped from every animal and every man, woman and child in the blasted town."

The drifter had never spoken in public before. In fact, he hid in the shadows and stayed out of view most time; that was best when Ma was drunk. He spoke in proper formal English now, like she'd taught him, rather than his daily vulgar slang, and his listeners thought him a fine orator. His strong voice carried his strong character, and his words conveyed conviction to his audience. He felt good and confident as he spoke and wondered how it happened.

Earl turned his head left, then right, then center to engage his audience and said, "Every indication is that Coyote Gulch is the next menu item for the wasters. Death in black, and it's either us or them. Deputy Palmer, now acting Sheriff, will address our proposed plan of battle."

Before relinquishing the podium, Earl looked straight at Sarah Peale and said, "I can only speculate that if we kill the evil peddler Kayne, even if we lose to the wasters, in the end, the monsters will eventually starve to death, for not knowing where to go next." Sarah looked down at her fretful hands and willed them to behave. She steadied her nerves and looked up to see Earl smirking as if victorious. She bravely gazed into his eyes and firmly nodded her head in agreement; kill him.

Acting Sheriff Palmer took the dais and said, "We have a lot of work to do. I want this Meeting House fortified tomorrow because the children, the sick and the elderly will take residence inside during the battle. We should be able to get two hundred people in here, with the second floor, and get the pews out of here. Board the windows, and side doors, and brick up the fireplaces and stoves!

"Next, we will need to stockpile some food and water inside in case of a siege or something. I want all the men to clean their shotguns and load them with buckshot." He pointed to a pudgy man with a broad-brimmed hat and said, "Thank you, Buck's Hardware, for donating guns and shells. We will post men at all three entry roads, set up barricades, and defend the town. Now I know it's gonna be hot, but we all must wear thick, protective layers of heavy clothing and any other measures of self-protection you can dream of. We don't have...."

Palmer was interrupted in his speech by the dramatic entrance of Red Sky and Danny, who held onto the fringes of the Navajo's deerskin shirt. The Sheriff faced the door and watched Red Sky, head and lance held high, his back straighter than an arrow, slowly stride up the crowded center aisle. Men coughed and grimaced but parted to let the two pass by. Everyone twisted around to see who or what interrupted their meeting.

Sibilant whispers and hissing murmurs leaked from sealed lips all up and down the rows; people didn't appreciate the Indian's presence. This was white man's business, and the old red-skinned, deluded fool should go home to the reservation and leave them to it. And why did the heathen bring the mute orphan boy along?

The acting Sheriff regained his composure to inquire, "What is the meaning of this Red Sky? You have no business here."

Red Sky stopped halfway down the aisle and tapped the butt of his lance against the wooden floor four times. Then he addressed the question, saying, "The meaning will soon become apparent." His deep clear voice, and excellent command of English, caused many to listen. He continued, saying. "Your battle will be fought in this world, and its grandfather will fight in the spirit world. Life Bringer and Death have waged war since the beginning and will fight until the end of all things. This is one battle, yet we must prevail, for the Life Bringer tests your will and courage."

Earl stood up and said, "What does the boy have to do with all this superstitious nonsense?"

Red Sky ignored the barb, reached behind him and took hold of Danny's small hand. He pulled the lad forward so that he stood in front of the Indian. Red Sky said, "Danny knows the images, voices, and scents of the spirit world. He understands the meanings and knows much that we do not. Ask him."

Palmer stood in place at the podium, letting Earl handle this disturbance for now. Earl smiled and waved the boy forward. "Come up here, Danny." The boy paused, momentarily torn, looking to Red Sky for assurance. The man nodded, pushed him forward gently and said, "Go, Danny."

Danny stared straight ahead, afraid to look left or right and see all the people ogling him, as he walked towards the head of the Meeting House. He joined Earl, who loudly asked the boy a question, one meant to discredit the dumb kid without overly embarrassing him; "Danny, do you know how to defeat the wasters?" The Indian was the real problem here, not the silly kid.

Danny shrugged his shoulders, and Earl smiled triumphantly, but Danny wasn't done. He mimed writing, and Dr. Ford found some paper and pencils inside a nearby desk. The boy sat at that desk, chewed on his tongue and printed five large letters. When finished, he handed the sheet of paper to Earl, who took it to the podium to read aloud.

His eyebrows rose as he read off the five letters, "w-i-s-k-y; whisky?"

Most of the literate focused on the five letters, not the sound of the word, and could make no sense of it; they were just a dumb kid, after all. Dr. Ford reacted differently, however, after a moment of reflection. He popped out of his chair as if catapulted. "Of course!" he yelled, "Of Course! whiskey! The drunkard Charlie survived, and the waster in the whiskey glass died!"

Earl saw it now too, but the realization dawned on the crowd slowly, and when the awful consequences of using whiskey as a weapon struck, it landed like a sledge hammer to the heads of many. Shouts of protests denied its possibility. One old timer yelled, "whiskey couldn't harm a fly. It's the best thing for you sometimes!" Another shouted, "I ain't wastin' my whiskey on no wasters!" People stomped and raised their fists, and Doc Ford needed several minutes and shouts of calm to regain control of the meeting.

Once a semblance of sanity took hold, Dr. Franklin Ford addressed the assembled again. "I agree with these noble sentiments, and before you folks get all maudlin and weepy about the wonders of whiskey, we must consider the greater good."

"Boo! Hiss!" His message met its match.

Red Sky spoke now from the center aisle, and heads turned to face the Indian. He said, "The waster army has already departed the Carson Ranch and will arrive in two nights. I will build and make sacred a sweat lodge to cleanse myself and offer to the Bringer of Life. You should also prepare, for Death comes soon on a million legs."

Bedlam let loose! Everyone stood now, and madness seized their minds. People pushed and trampled, shoved and shouted to escape the confines of the Meeting House as if suddenly infected with mass claustrophobia. The building emptied more quickly than a drink in Ready's quick hands, and most of the exodus joined him at the Cactus Juice Saloon. Others stopped in at Moonie's Dance Hall or Madame Clara's place. If they only had two more days to live, it was time to do some fast living.

Several remained in the red brick building, including Earl, Dr. Ford, Sheriff Palmer, the parson, Danny and Sarah Peale; Red Sky had departed. Earl thought that Sarah looked disoriented and unsure, for she stood in the aisle facing the exit but didn't move. A sadistic urge overcame the drifter, and he hollered out, "April?"

As she had done for twenty-two years, Sarah Peale responded to the name April, and she turned to face Earl before realizing her error. She choked, covered her face and ran out of the building. She stumbled and fell at the threshold but got to her feet and continued her flight to the boarding house and her room.

Dr. Ford frowned and said, "That was just plain cruel, Earl. You have no heart at all; yer just hard and unbreakable; unloved and unlovable, like cold steel." He shook his head and stamped out of the building. Earl did not comment. He left, too and went into the Cactus Juice Saloon. The Sheriff returned to his office to look at maps and study defense strategies.

Later that night, Earl awoke in his bed to the sound of many hushed voices and heard people moving about. Curious, he pulled on his trousers and went into the hallway barefoot, careful not to awaken Danny. A lantern was burning, Mrs. Cally was up and about in the kitchen, and then Earl saw Dr. Ford exit the room two doors down; Sarah's room.

The doctor acknowledged Earl with a shake of the head and explained why he was there at two in the morning. "Sarah miscarried."

Shock and dismay colored Earl's face gray, and he said," Oh my God! Her father's?" The drifter had jumped to the worst possible conclusion because life was like that for him. The doctor said, "No, a man who cut wood for the miners. Her father and his clickets killed him. That's what the peddler calls the wasters; clickets. She lost a lot of blood, but I think she'll be okay now after a couple of days. Just needs lots of bedrest and liquids. In fact, I'm going home for some bed rest myself. Got two big busy days ahead, maybe my last days ever."

Earl made to go into Sarah's room, but Ford said, "Wait until tomorrow; I gave her something to relax. Good night Earl." Danny, peeking out the door, saw and heard it all, and he jumped back into bed and feigned sleep.

The widow Cally said, "Thank you, Doc, that's my prescription too; sleep. Good night."

Earl also retired to his bed, happy to see Danny sound asleep. Then, a couple of hours later, Earl awoke again, roused by another voice! He thought he was dreaming, but the whisper insisted that it was real; Sarah's voice. Earl opened his eyes and saw her as a darker part of the darkness of his room, standing near his bed. He sat up quickly, rattled and confused. Danny stirred slightly, moving his head so he could hear better, and pretended to slumber.

Sarah laughed lightly at Earl's apparent discomfort and

whispered, “You’re just so damn smart, aren’t you? “Acidic venom dripped from her words; her tone of voice burned his ears.

He answered softly, “Not so much with women. I ain’t been socialized much.”

“How did you know me? I’ve been so careful to hide from my past!” She sighed, exasperated.

Earl sat up in bed, propped against the headboard. He struck a match and lit a candle on his bedside table, and once illuminated, he said, “Fifteen years ago, you and your Pa came by every once in a while; my Ma was single too then. I remember you had to stay up on the wagon while your pa and my ma…; well, you weren’t any older than four or five, but those green eyes of yours gave you away. I wasn’t sure at first, but….” Danny rolled his eyeballs under his lids but remained motionless and breathed steadily.

Sarah sat on the edge of the bed near Earl because standing caused more severe abdominal pains. She said, “My mother ran off when I was a toddler, went out to San Francisco with some lawyer; Pa always said she was a whore. We moved out here from Missouri when I was six or seven, and he began his trading routes, but then I didn’t ride along anymore when I got older, so I could keep house and milk the goat and care for the chickens and such.”

Earl wasn’t certain why. Maybe he felt he owed Sarah some explanation too, or maybe it was because his brain still slept, but he told her the truth about his father, something he'd never admitted out loud or faced up to before. He said, "I always told folks that my Pa died a hero, fighting for the south in the Civil War, but the real story was that he died a coward, executed for desertion." The shock of saying and hearing those words seemed to break something inside of Earl, it felt like a dam broke, and pent-up feelings held back for years suddenly gushed out.

Sarah was less astonished than Earl by the teardrops that rolled down the man's cheeks. She wiped one away with a flick of a tender finger. "I knew there was a human being in there somewhere," she said as she stood to leave.

Earl said, "That's more than I knew. Thank you." Then, like an idiot, he asked, "Did you love him?"

"Good night Earl." Sarah gently closed the door behind her and walked to her room, where she lay on the bed and soon fell fast asleep, wondering if she ever loved Foster.

Chapter 8

Morning awoke to a hammer concert. A crew of early risers nailed planks across the Meeting House windows and side doors, attaching boards to the inner walls. Children created a bucket brigade to stockpile water supply for tomorrow night's captives. A few carpenters constructed an indoor toilet facility behind a curtain, and masons bricked up the fireplace and plugged the stovepipes.

The personal business was done yesterday, and people worked together today; the heavenly scent of freshly baked bread and pies arose from many a kitchen, and women cooked up great pots of food for the laborers.

Sheriff Palmer appointed Earl his deputy, as recommended by Dr. Ford, and then assigned sentry duties to men armed with shotguns. Three teams of four men each, two men per shift, would guard one of the three likely entry points into town, six hours on and six hours off.

Two rival factions formed in the town, at complete odds with each other. One group insisted that whiskey was harmless and shouldn't be wasted on trying to kill the monsters, while the opposition made plans on how to distribute and apply the liquid weapon. At one point, people heated up to near boiling, steam rising from their ears when Sheriff Palmer officially impounded the three and a half barrels donated by the town's saloons.

Barkeep Shultz assisted the Sheriff, and neither felt safe until the precious, liquid ammunition was safely locked up in the town's only jail cell. They estimated their haul at about two hundred gallons. The opposing sides fussed and fought about it anyway until the saloons announced free drinks until midnight, when it would be against the law to imbibe. Then the adversaries went to the saloon to argue over drinks together.

Ready danced atop cloud nine points nine; he became the center of attention at the Cactus Juice Saloon all day long and preached a drunken sermon to the faithful. He slurred, "Shust be sure ya is plastered drunk when the critters come and yewel be jist fine. Look, if'n yous can walk or stand, you ain't drunk enough; have another botto. Thas the onliess way."

The whiskey preacher's flock believed his gospel and partook earnestly like a sacred rite. Charlie Ready still owned enough coins to maintain his chronic drunk. "And besides," he cheered on his disciples, "ya gotta drink until your piss is rye."

Before noon that day, Earl and the three wranglers from the Carson Ranch collected the townsfolk's many horses and mules, plus nearly a hundred and twenty women and older boys and girls as riders, and led them twenty miles out of town into the eastern foothills. Once in a suitable area of high ground, folks worked together to erect shelters and makeshift kitchens. Guard duties and horse care were also assigned to the older boys and girls.

Early evening, Earl and the other men finally rode back into Coyote Gulch. Earl put up his horse and went directly into the boarding house, where he sat down alone to a bowl of cold stew for dinner. Afterward, he visited Sarah. He knocked on her door and announced himself. She bade him enter. Danny sat with her, communicating in some kind of hand language and gestures. The boy's face lit up when he saw the dusty, tired cowboy who'd rescued him, and he ran to greet Earl with a big hug around his waist.

The man of cold steel blushed, grinned sheepishly, and said, "Howdy, ya'll." He patted the boy on the back and looked at Sarah. He asked her, "How are you feeling? I mean, well, I don't know what I mean, but does it hurt more physically or emotionally?" Gawd, he felt so stupid sometimes!

Sarah said, "I'm sick inside and out. You know, it sounds crazy now, but I wanted to rush into a marriage when I knew I was pregnant to avoid being stigmatized, but now it all seems so cynical and pathetic of me. I didn't love that man; perhaps I am a product of my rotten father after all and a whore like my mother too."

Earl collapsed to his knees at her bedside and grasped her limp hand. He pressed it to his lips and said, "And I am a coward like my father and an abuser like my ma. I don't know nothin' about love, Sarah, but woman, you make me want to know all about it."

He didn't know where the words or strange emotions came from and wasn't interested in learning right now. There were priorities.

He'd had women before, many times, but they were for fun and games, nothing serious; life was that way for Earl. Now, suddenly exposed to mass death and a real sense of his own mortality, Earl needed an anchor, something weighty to hold onto so he didn't get washed away by the flow of events. It seemed he couldn't control everything, that life happened despite trying to avoid it.

Sarah said, "Well, I'll be a plucked chicken Earl, but love isn't a subject I'm learned in either. It's just something I heard about."

Danny raised a hand; he knew the answer. Earl and Sarah smiled at the eavesdropping boy, and then he mimed a kiss, puckering his lips and closing his eyes.

Earl laughed it off and said, "Well, there's another big meeting tonight to finalize plans for tomorrow. I'm certain that Doc Ford will prescribe for you to stay safe inside our fortress there tomorrow. Good night." He departed, and Sarah indicated to Danny that she needed to rest.

It was nearly seven and ninety-some degrees when Deputy Earl Gage stopped for a moment in front of the wigwam that Red Sky erected for a sweat lodge. He hadn't yet decided; whether the Navajo was a crazy heathen or perhaps knew things other men had forgotten. Steam rose from a hole in the top of the teepee, and a bucket of water sat outside the closed entry flaps. He shook his head and moved on.

The double doors to the Meeting House were wide open, and Earl strolled in. Sixty or more men stood inside, but it looked near empty tonight, with the benches removed and furniture stowed. Last night, doors and windows allowed fresh air and light into the building. Tonight it was dark, stuffy and sweltering inside. Candles and a few oil lamps lent a dim dancing light to the assembled men, perhaps sixty in all. Blankets and pillows lined the interior walls now, and water buckets stood in each corner.

Earl's boots clacking on the wooden floor echoed in the chamber. The cloying stench of unwashed bodies filled his nostrils, and he almost gagged but guessed he smelled as bad. Earl found a place to plant his feet and saw the doctor's faded brown hat and gray-haired head bobbing several men rows ahead of him.

Sheriff Palmer spoke alone tonight. He stood at the podium and succinctly described a detailed plan, assembled with the assist of Deputy Gage, Dr. Ford, barkeep Shultz and others. "I thank you all for coming," he said. "We all know and understand that we are dealing with serious life and death issues, so starting at midnight, it will be illegal to drink whiskey. If booze kills these wasters, we must conserve all we can! It's our secondary weapon. Shotguns come first."

He continued: "Our forces will be divided into three squads to defend the three roads, the most likely enemy's entry points. I will lead one crew of twelve, which will blockade the intersection of Main Street and Raven Road, the most obvious ingress from the Carson Ranch, while Deputy Gage will defend the disused trail leading northeast to Bleaker and the mining camp. Dr. Ford and his eleven volunteers will cover the wagon trail leading to destinations south of Coyote Gulch.

"Every team member will be armed with two double barrel shotguns, loaded with buckshot, and one wooden bucket filled with whiskey. Each of the three posts will have a nearly full barrel of the stuff to supply the buckets. We shoot first, in unison, to blast away their front lines; both barrels, both guns! Boom! Boom! When and if the multitude advances, we resort to the whiskey for close-in fighting, while any extra volunteers can reload the shotguns. So shoot at first sight, and splash when near.

"Now, about staying safe out there tomorrow night; first, no fires. Secondly, no drinking! And third, dress for the occasion; wear heavy layers of thick leather, and cover your arms, legs and head if possible. Wrap a wool scarf around your head, whatever, to protect yourself. Now go home and be with your loved ones. This may be your last night to do so." The building emptied quickly to escape the odorous oven and to do as the Sheriff suggested. Somber, sober men went home to their families.

Earl visited Sarah in her room again later, though Doc Ford assured him that she was recovering from her severe blood loss and her baby loss too. Gage knocked and entered. A burning oil lamp flickered from a table, and Earl noticed how pale Sarah's usually rosy cheeks looked. She smiled brightly, however, and sat up in her bed. "Earl; what brings you here?' she asked.

"Hi," he replied lamely, doffing his hat and blushing at what brought him here, but he said, "You did, Sarah. It's weird now, but when I was twelve or whatever and you and your pa stopped in, and you were only five or six, I always said to myself, "I'm gonna marry that girl one day." His face turned so beet red at the admission Sarah laughed at his embarrassment.

She said, "That's the sweetest lie I ever heard, Earl."

That response flummoxed and flustered Earl to such an extent that he was near tearing his poor, abused hat into ragged little pieces. Sarah patted the side of her bed and said, "Sit down here."

Earl assumed a defiant attitude and said, "It's true, and I ain't sittin' there 'til you apologize."

"Oh, don't be so sensitive and sit down. Okay, so I'm sorry to doubt such a paragon of honesty."

Earl stifled the urge to kick the bed, and docile as a lamb, he sat on her bed. More color pinked Sarah's cheeks now, and Earl desired to kiss her badly. He was wrong for her, however, or for any real woman for that matter. He was undomesticated, like a feral cat, and no good to anyone, especially not for Sarah, Danny, or Coyote Gulch. The very idea of Deputy Gage was laughable. What the Hell was he thinking?

Sarah asked, "So now what? Do we all die tomorrow from the peddler's death, or can whiskey save the world?"

She was a sassy one, but Earl liked it, needed it even. He said, "I can't answer that, Sarah. Red Sky would say it's in the hands of the Life Bringer, but the guns and whiskey buckets will be in our hands, so I just don't know. At least you, some other women, and children will be safe inside the Meeting House tomorrow night."

Sarah said nothing, so Earl said, "I'll be real busy all day tomorrow, setting things up and all, so I better be getting to bed now." He stood, still fiddling with his hat, and faced her with saying good night, but she spoke first, saying, "Wait! Before you go, do me a favor; since this could be our final night on Earth. It seems only appropriate that we should kiss as if it's the last one we will ever have. So put down that stupid hat and kiss me."

Earl needed no special pleading to comply with such a request, and he obeyed.

Chapter 9

The morning's red sky bodes ill, and a palpable sense of fear gripped the entire community of Coyote Gulch, for this could very well be its last morning on Earth. Fluffy, puffy clouds slid slowly across the crimson dawn sky in columns, like a navy of mashed potatoes sailing forth. Red Sky also noticed the omen when he fetched water to pour over the heated stones in his sweat lodge, scowled at the sky, and returned to the tepee. Danny saw the sign and smiled.

The Sheriff, the doctor and Deputy Gage chose their teams and set to work, wheeling the whiskey barrels to their assigned posts and stockpiling shotguns, shells, and wooden buckets. Some of the men built walls of rocks or boards and logs as barricades for some measure of protection from the wasters.

The entire community united in their efforts, which continued for much of the day, their labors only broken by a fabulous noontime dinner served up by dozens of women and girls, including the late Sheriff's two teenage daughters. Beef steaks, fresh bread, biscuits, potatoes, and corn on the cob, were served, with apple pies for desert. Liquid refreshments consisted of beer, coffee and water, but no whiskey. It was illegal to drink today, and severe penalties for violators of the Sheriff's ordinance were threatened. Danny stuffed his pockets with food for Red Sky.

After the big dinner, the children, the sick and the elderly were shepherded or assisted into the fortified, red brick Meeting House. Dr. Ford ordered Sarah to accompany the groups, and he settled her into a cot on the second floor, though she protested the special treatment. The doctor cocked his head to one side and said, "There's nothing special about it, Miss Sarah Peale, but I am starting to believe there is something special about you." He winked.

When they learned of her relationship with the so-called peddler of death, Sarah never expected such kind understanding from people. She said, "But I'm tainted, aren't I, Doc? Related to a maniac and pregnant out of wedlock too; stained, marked." She buried her face in her hands, but she'd run out of tears.

Ford nodded his head and replied, "Yes, in your own mind, you bear the sins of your father, and you lost the baby as divine punishment for your sins. I don't believe it for one second, young lady! Not for one second, and most decent folks around here don't believe it either, including Earl, it seems."

"What do you mean?" she asked indignantly, though her cheeks pinked.

The doctor smiled and said, "It's no secret how he feels. He talks about you all the time."

Sarah's heart jumped, and she told the doctor, "That's the sweetest lie I ever heard, Dr. Ford." He got a chuckle out of that, and then he said, “Well, you behave now, ‘cuz I’m off to the wars.”

“Good luck Dr. Ford, and thank you for everything.”

Charlie Ready took a nap on the haystack in Potter’s Livery and Stables. Yesterday, he’d accidentally discovered the two cases of whiskey under the hay, whiskey that Earl had hidden there but had since forgotten about, and Ready was a happy drunk all over again. It was illegal today, but no one knew of it but him. He loved this town.

General Jethro Kayne led his three hundred thousand clickets south, just off to one side of Raven Road. He and his army traveled slowly because they ate as they advanced. Every grass blade, weed leaf, flower, cactus, insect and vole became fodder to march on. The eaters and marchers traded places constantly, at about one or two miles per hour at night, and rested in the shade daily. When those in front encountered edibles and stopped to dine, the clickets behind hopped or climbed over them and so on so that the hungriest clickets would always be on the front line of the advance. The infernal racket of their clicking, crunching mandibles continuously accompanied their noisome advance.

This dark black night, with moonlight and star shine blocked by tall, thick piles of bleak, threatening clouds, was perfect for an attack but an obstacle to travel. Kayne couldn’t see his own feet! Progress proved slow and difficult for him, but the clickets had no such issues. They preferred the camouflage of darkness for their mass attacks. The Kwarlik army traveled in a block formation, the individuals remaining close together; front, rear and sides; shaped like a square, and from two hundred fifty to three hundred feet wide and deep.

Kayne knew they were making bad time, but it was much too dark to read what time it was on Mr. Carson's gold pocket watch and know how long they'd been on the move. He didn't know where they were for certain either; even if he might have remembered some landmarks from his years peddling this route, he couldn't see them now. He was lost in time and space; he didn't know how far they'd come nor how far they had left to travel. He simply followed alongside Raven Road and hoped for the best. The clickets seemed to sense when they somehow closed in on their prey because they always went silent, so there was that. Qweequa rode in his vest pocket and depended on Qweequa, the queen clicket, to warn him when near town.

Chapter 10

The stress and strain on the town's defenders grew more potent by the hour, and after one o'clock in the morning, the waiting for death became nearly unbearable for some. Earl wanted to scream and pierce the tension like a lightning bolt rents the sky. Hours earlier, between eight and midnight, Earl and the other men joked and made light of their duties, using black humor to combat the fear of the black, clicking death.

However, the mood transformed into a tense silence after three or four hours when even the sound of breathing was frowned upon. Nerves frayed, and men unraveled like cheap twine; fearful thoughts spun new yarns, colored blood red and black. The men tried to maintain a calm confidence they didn't feel; no one wanted to appear weak or cowardly. Inside they seethed and boiled, threatening a furious explosion if the pressure grew any greater. The men sweated buckets, which further irritated them. Their extra clothing layers, coupled with a hot, humid night, and looming death, made for evil tempers.

It was the darkest night ever known to boot, and the impenetrable blackness was only accentuated by the occasional, blinding lightning flash, followed by a dog growl thunder bark, which scared the men even more than the waiting. It felt like God was nearby, and they weren't sure which side He was on.

Raven Road spanned a width of thirty-five feet, give or take, and Sheriff Palmer, plus eleven other shooters, spread across it, quaking with apprehension, but acting macho, behind a short wall of rocks and stones. The barrier also sheltered thirteen wooden buckets brimming with whiskey, twenty-four double-barreled shotguns, and one hundred extra shells, with four volunteers to reload the guns and buckets. The Sheriff and the half-full whiskey barrel occupied the center to facilitate bucket refills.

A dozen shooters stood between a hotel on their right side and a café to the left. The only way past the defenders would require a detour around the businesses and through narrow alleys, where two armed men stood guard, outfitted with four guns and four buckets of whiskey.

Red Sky emerged into the night from his sweat lodge after twenty-four hours of fasting, praying, and cleansing, and he felt faint. He squatted down for a moment until his dizzy spell passed; he had a few minutes to spare. After regaining his equilibrium, the unperturbable Navajo strode through the deserted town square towards Sheriff Palmer's squad on Raven Road and Main Street. He carried the sacred lance tipped with a single eagle feather.

One of the sentries felt the Indian's silent approach and nearly spoke, but Red Sky hissed, "Shhh!" Sheriff Palmer heard and saw the Indian, as a shape darker than the surroundings, and he whispered, “Get out of here! Go back to yer wigwam.”

Red Sky sternly but quietly replied, “Please believe me now, Sheriff, and do as I say. Aim your shotgun chest high at one o’clock, and use both barrels. Shoot when I say now.”

The Indian couldn’t see any of the men’s expressions of disbelief and confusion and didn’t care. After a long deep breath, Red Sky whispered, “Prepare,” He pointed to one o’clock and soon shouted, “Now!” Simultaneously, Danny, who sat with Sarah on the second floor of the Meeting House, jumped up and silently screamed, his mouth open wide, sans sound, startling Sarah. She couldn’t calm the boy for several minutes and wondered what he’d seen or heard that she could not. Then the shooting started.

Unfortunately, the Sheriff hesitated briefly so that when he did aim his shotgun and pull the triggers, he only blew off General Kayne’s left arm at the elbow and didn’t kill him as he might have otherwise done.

Kayne’s instinctual howl of shock and pain signaled that the battle was on! Eleven men, each an arm’s length away from the next, shot their guns all at once, and twenty-two barrels blasted forth a fusillade of buckshot, killing hundreds of the black attackers. Kayne had dashed for the safety of the restaurant’s back wall after losing half of his left arm and was not hit a second time. When the lightning flashed, he saw the splintered bone and lots of bleeding where his wrist and hand used to be attached.

Jeth used a dirty bandanna and his teeth to tie a fairly tight tourniquet above his elbow, and the blood flow slowed considerably. Dammit! He hadn't heeded the clickets going quiet! And why hadn't Qweequa warned him; her safety depended on his! If only he'd been able to see how close he'd gotten to town! He could have stayed back like usual and avoided this wound, but he truly did feel like a real soldier now and finally understood the agonizing glory of war.

As soon as Dr. Ford and Deputy Gage heard the first volley, they abandoned their posts as per instructions, and their men wheeled the whiskey barrels and carried the buckets and guns to Palmer's assistance. Palmer's men swiftly unloaded their second shotguns as well while volunteers reloaded as rapidly as possible. It wasn't fast enough, though, and the ear-splitting cacophony of clicking grew louder and closer despite their buckshot.

The deafening clicks struck fear in the brave men, but they held strong, and when the Sheriff hollered, "Splash," the men grabbed their whiskey buckets and tossed the contents over the wall and onto the streaming, clicking hordes. Thousands of expiring sighs escaped the dying wasters. The booze worked! It killed them!

Thousands more of the attackers hopped over their dead, and the men were suddenly under assault around or between the splashes! The wasters spilled over their wall of stones and were upon them! The Kwarliks walked through and over the whiskey without any effect, but if the toxic liquor washed over them and saturated their breathing pores, it absorbed into their tissues and killed them almost instantly.

The Sheriff and Red Sky rapidly dunked pails into the open whiskey barrel to fill them and tossed the contents over the screaming men, killing the clothes-tearing, flesh-ripping little black monsters of the night. They poured buckets over themselves and killed all the clickets clinging to them, but the next wave of attackers swept in right behind.

Before the Sheriff went down under the onslaught, he pushed the whole barrel of whiskey over, and gallons splashed and spewed out, wasting the wasters in its path. Red Sky never dropped his lance as he danced and hopped about like the wasters, stomping on them and shaking many off as well. He couldn't succumb yet. One task remained to him.

Earl and company arrived on the scene only seconds after Dr. Ford's men and whiskey had, and soon thereafter, the precious booze splashed everywhere. The men poured buckets full of whiskey over their heads and soaked down those men under direct attack, killing many thousands of clickets and holding back hundreds more.

The four men holding the alleys between the buildings bracketing Raven Road were attacked and soon overwhelmed, allowing a black wave to flow through each passage, and now the defenders of Raven Road and Main Street were flanked; surrounded even, by the rapacious wasters, and running out of whiskey.

Jethro Kayne, sitting up against the restaurant's back wall, knew the shooting was over; his clickets were winning the battle. Qweequa peeked out of his pocket and clicked a message about the fight, something about whiskey, but Kayne couldn't decode the words as quickly as Qweequa clicked them. And besides, he was angry, and though fearful of making an enemy of the queen, Jethro accused it of betrayal. "You let me get shot! You didn't warn me. It has coulda been you, ya know!"

Qweequa took the risk. After a victory here, they wouldn't need Kayne anymore. Qweequa knew her way around now, and she'd guide the hordes. She clicked a reply, but Kayne ignored the critter; he had things to do. He moved out towards the conquered alleyway. He picked up one of the dead sentry's shotguns in his remaining right hand and jogged toward a big brick building he'd seen. Lightning flashed more frequently now and lit a path for him. He guessed correctly that the children and women hid there and were the key to winning this town.

Charlie Ready woke up and wondered what the confounded racket was all about; non-stop shooting, a horrendous clicking sound, and sometimes thunder, like Hell's orchestra! He rose from the haystack he'd napped in, checked to ensure that a case and a half of whiskey remained beneath it, and smiled at the lovely sight. He gulped a few ounces of the life-giving fluid, peed, yawned and stretched before staggering outside to investigate the cause of such inappropriate commotions. It was the middle of the night, for land's sake!

Ready wobbled and weaved unsteadily in the center of the empty blackness of Main Street, fighting upstream against the noises until he found the awful racket's source. A bright bolt of lightning illuminated the night, and to Ready's bleary eyes appeared a large group of nearly fifty men trapped inside a shrinking circle, enveloped by those horrible clicking, biting bugs! He wondered if he was dreaming about what he'd witnessed in the miner's camp. He pinched his cheek, and it hurt, so this wasn't a nightmare. It was real. He'd done this once already and wanted nothing to do with it! A powerful thunder cracked the sky as if trying to split the heavens apart.

His curiosity cured, and not wishing to be chewed on by the clicker bugs, Ready turned to go back to his cache of whiskey under the haystack but was stopped by a shout. Red Sky called, "Ready! Catch!" The besieged Indian, standing against the clicket onslaught but soon to be buried beneath the blackness, threw his lance, and without thinking, like usual, Charlie Ready tried to catch the sacred weapon. His timing was off, however, and he missed, but the six-foot-long missile stuck harmlessly into the dirt between the drunkard's feet.

Charlie Ready looked at the quivering spear dumbly for a second, wondering where it came from, and absently heard Red Sky yell, "Plant it in the cemetery! Now." Lightning flashed, immediately followed by an explosive thunderclap, and when Ready's night vision cleared, it seemed all the encircled people had vanished beneath a clicking, storm-tossed, black sea.

The drunk pulled the lance up and processed the Indian's request; something about planting the spear in the cemetery, like it was dead or something. Did it need a coffin? Or maybe the Indian thought it would grow. Ready owed Red Sky for leading him to Coyote Gulch, but he wasn't diggin' no six feet deep to plant a spear. That'd be crazy. The good deed doer wasn't quite certain which way was which until a lightning bolt lit up the Meeting House. The cemetery was up the hill, right behind it. He didn't understand the request, but it seemed a kind thing to do while he was out here anyway, and Red Sky seemed busy, so Charlie Ready walked up the dark hill to the graveyard.

Meanwhile, Jethro Kayne had reached the Meeting House and stopped at the doors to carefully consider his next moves. The door was sure to be barred, and he would have to trick the inmates into opening it. He didn't want to set his shotgun down; he held the stock against his chest with one hand, his finger resting against the trigger guard. He'd have to kick the door softly, on the lower half, like a child might knock, and wait.

Just as the peddler raised his foot, Danny alerted Sarah. He popped his lips, and Sarah heard' "P! P!" And then the mute pretended to raise and aim a shotgun. Sarah didn't understand, and then she heard a thudding from downstairs as if someone was kicking the door. Sarah asked Danny, "P? Like a gunshot, pow pow?"

No, he shook his blonde head side to side, popped his lips again, and exhaled hard, and Sarah heard, "Pa, Pa!"

"Oh my God! He's at the door!"

Yes! Danny nodded furiously.

Sarah acted instantly. She rose from the cot and ran down the stairs barefoot. She saw a young man with a shotgun looking worriedly toward the front door, and Sarah grabbed the weapon out of his hands before he knew she was there. Danny kept close to her, flying heels. The frightened children down here on the first floor, and the older girls and women who'd watched over them, huddled together in a tight mass, as far from the threatening door as they could squeeze themselves.

Sarah barked to Danny, “Open it!” She brought the gun up to her right shoulder and rested her finger on the triggers, prepared to blast both barrels into her father, the peddler of death. Danny lifted the iron bar from its holds with difficulty, managed it, then dropped it to the floor with a loud clang and dove to the ground close to the wall.

To open, the doors had to be pushed or pulled outwards, and Sarah wondered why her father didn’t yank one open, so without hesitation, she shot through the center of the double doors. They flew open, and Jethro Kayne stood unharmed by her efforts. The thick oak doors absorbed the full fury of the blasts. However, Pa’s shotgun pointed at her chest now, and he laughed. “Ha ha, I lost an arm but regained my dear, sweet, loyal daughter; what joy.”

By this time, Charlie Ready arrived in the graveyard and set the spear down on the hard ground and kicked some dirt over it. Then, good deed done, he longed for a drink to celebrate, so he went to the only place in town with whiskey, his private stash under the haystack at the stables.

Standing on the floor, Danny saw the Life Bringer in his mind again, another vision, and learned what had to be fixed to save the land from Death, and only he could do it. He stood up, back flat against the wall next to the open doors. The bad man hadn't seen him back there. All attention focused on Sarah, who he threatened and cajoled, like a cat with a mouse.

Danny dashed from his hiding place and pushed past the peddler, bashing into the man's bloody arm stub, and then the boy ran as fast as he could towards the cemetery. Jethro Kayne roared in angry surprise and turned to fire at the fleeing brat, thinking he had no worries at his back, for his unarmed hostages were cowed and completely in his power.

Sarah wasn't. She lowered a shoulder and ran forward to slam into her father's back and punch a fist into his wounded half-arm. Jethro yelped and sprawled forward. Kayne spun to the right as he tumbled and tossed the gun, hoping to break the impact of his fall with the freed hand and avoid landing on his injury. Sarah's momentum carried her out the door, but she kept her balance. Without a second thought, she snatched up the abandoned shotgun, took a stance, and blew her father's head off.

Danny, in the cemetery, uncovered the spear, picked it up, and plunged the point into the dirt with all his strength! Multiple blasts of lightning flashed, and ear-shattering thunderclaps instantly answered, cracking the sky. Clouds burst open, and Danny danced for joy. The Life Bringer had answered!

Only seven minutes had elapsed since Red Sky tossed the lance to Ready, but several defenders were already dead while the living brutally battled for their lives, bloody and torn, rolling towards puddles of whiskey or whatever means they could employ to kill the wasters. The man's thick, layered clothing only protected them for so long, and some of the affected men ran this way and that, in absolute painful terror, clickets clinging to their faces and hands, eating them alive!

Before the downpour began, the deafening clicks seemed to announce victory, but it wasn't water that now cascaded from the sky like a cataract. It was aged bourbon whiskey! Red Sky, bitten, blinded, and dying, tasted the booze, which had never seemed so sweet. The clicking, biting, hopping, and crawling ceased, and great sighs rose from the dying wasters; they'd been wasted. The battle was won; the Bringer of Life conquered Death this time around, but not for Red Sky, Sheriff Palmer, barkeep Shultz or eleven other brave men.

The wounded stopped in their tracks, rose to their feet in utter amazement, opened their mouths to the delicious blessing from the sky, and shook the dead wasters from their clothes and bodies. Coyote Gulch was saved!

Dr. Ford and Earl Gage ignored their painful, bleeding sores and ran for the Meeting House; they'd heard shots! The cloudburst of whiskey from above gradually turned into pure water and slowed to a drizzle as the two men arrived there. Sarah stood with her back pressed against a red brick wall near the front door, where lay the dead body of a mostly headless man. Sarah gripped a shotgun in both hands, her teeth clenched tightly, and her eyes vacant, but she lived! Earl ran to her and pulled the gun from her hands. She resisted for a moment, but her eyes focused, recognizing Earl. Sarah gave up the gun and her arms to Earl, hugging the man while weeping, like when she'd lost her mother those many years ago.

Other wounded defenders limped up to the Meeting House too now, and several happy families reunited, and many tears were shed. Mournful howls rose from the new widows. Danny returned from his heroic act in the cemetery, sad and disconsolate; Red Sky was dead. Danny smiled, however, when he saw Sarah and Earl hugging and kissing. He ran to Sarah's side, overjoyed that she lived, and joined in a group hug with Earl and Sarah.

The people celebrated the wonderful blessing of the whiskey precipitation without daring to question its source, giving all thanks to the Lord. While the townsfolk looked to Heaven or hugged each other, Qweequa plotted in the dry safety of the dead peddler's upper vest pocket. She was a Progenitor and carried the seeds for a new colony inside.

Although it was dark and water had fallen, washing away much of the toxic alcohol, Qweequa considered the risk of venturing out as too great; there were too many humans around. She'd bide her time; no one seemed in a hurry to dispose of the body. The vest pocket was safe; she'd depart for new territory before dawn.

Charlie Ready was inside the stables when the whiskey rained down from above, and he'd missed the entire episode, though he wondered at the sudden cessation of the clicking. The thirsty man would have been devastated to learn of losing out on the baptism from above. As it was, he felt all warm and fuzzy; his good deed for Red Sky had inspired Ready to further altruism, and when he arrived, staggering, at the Meeting House a short time later, he pushed a dolly ahead of him, loaded with the remaining nineteen bottles of whiskey from his stash.

It seemed that no one noticed him until he shouted, "Hey, is it legal to start drinking again?

Cheers greeted the new hero, and the survivors descended on Ready like buzzards on a carcass. Earl imbibed as well and didn't even grow angry when he realized it was the same whiskey he'd rescued from anonymity in Bleaker. It mattered not. He had a lot to celebrate; it seemed like he had a family, and he was sure to be named Sheriff. Earl didn't care to wonder how it all came to be; if not for the clicket invasion, he'd still be another anonymous dust mote drifting with the wind. It didn't pay to question good fortune or the Life Bringer either

Charlie Ready noticed the body lying on its back and wondered why the man wasn't drinking. He shouted, "Hey you! Wake up!" Bystanders thought he was joking, but Ready wobbled and stumbled closer to the peddler's decapitated corpse. He tottered dangerously and stared down at the body for a moment before he noticed the missing arm and head. Charlie exclaimed, "Sorry mate, I know just how you feel." And then, with a large audience watching, he lost his balance and fell over.

Ready dropped like a stone and crumpled directly on top of the cold dead peddler. His right knee crashed down hard onto the cadaver's top vest pocket. Ready never heard Qweequa's death sigh over the noise of his flopping about like a spastic turtle on its back as he tried to the right himself. Somehow, despite the injuries, suffering and deaths, everyone laughed hysterically.

THE END

About the Author

Mars Knight is a retired widower living in Belen, New Mexico, where you can wash the car, and it still won't rain. He is writing under apseudonym because he doesn't want to embarrass his family. Mars loves his four dogs, the great outdoors, and writing, but not necessarily always in that order.

Made in the USA
Monee, IL
04 November 2026

67382307R00066